BestMasters

Springer awards „BestMasters" to the best master's theses which have been completed at renowned universities in Germany, Austria, and Switzerland.

The studies received highest marks and were recommended for publication by supervisors. They address current issues from various fields of research in natural sciences, psychology, technology, and economics.

The series addresses practitioners as well as scientists and, in particular, offers guidance for early stage researchers.

David Roos Launchbury

Unsteady Turbulent Flow Modelling and Applications

Springer Vieweg

David Roos Launchbury
Horw, Switzerland

BestMasters
ISBN 978-3-658-11911-9   ISBN 978-3-658-11912-6 (eBook)
DOI 10.1007/978-3-658-11912-6

Library of Congress Control Number: 2015954647

Springer Vieweg
© Springer Fachmedien Wiesbaden 2016

Springer Vieweg is a brand of Springer Fachmedien Wiesbaden
Springer Fachmedien Wiesbaden is part of Springer Science+Business Media
(www.springer.com)

Acknowledgments

My first thanks go to my supervisors Dr. Luca Mangani and Dr. Ernesto Casartelli for supporting me throughout this thesis and for giving me carte blanche on themes and procedure. I would also like to thank the many other professors who gave me the knowledge that I have today. Most notably, I'd like to thank Dr. Thomas Staubli and Dr. Andreas Haselbacher for their inspiration and their ability to motivate a passion for the subject matter, as well as Thomas Tresch for sparking my interest in fluid dynamics in the first place and for being a good friend. Further thanks go to Dr. Giulio Romanelli for the many fruitful discussions and his patience when answering my endless array of questions.

I also thank my colleagues Oliver Ryan and Simon Roth for their company during my studies and for making the long evening hours much more interesting.

I would like to express my gratitude to my parents Robert and Eva for the moral and financial support over the course of my entire education and for the patience they showed when it came to me finding my way. I'm also very grateful for the support of my parents-in-law Bryan and Maggie, especially for their endless hours of babysitting.

Finally, and most importantly, I would like to thank my wife Susan and my boys Andreas and William for putting up with me during my studies and for being the most important things in my life.

Abstract

The present work deals with the improvement of a previously developed explicit third-order time-accurate solver for transient flow problems implemented in OpenFOAM. As such, the solver is enabled to solve the spatially filtered Navier-Stokes equations applied in large eddy simulations.

An optimised pressure-velocity coupling algorithm is implemented to reduce pressure oscillations at small time steps. The support for a temperature transport equation is added with the aim of solving problems involving heat transfer in incompressible flows. Momentum and temperature source terms are added to allow for periodic boundary conditions in such cases.

The solver is validated on a series of test cases involving the flow between parallel plates and around a square cylinder. The flow over a turbulator geometry involving heated walls is investigated, as well as a jet-in-crossflow setup of a film cooling case. In all these cases, the performance of the static Smagorinsky, dynamic Lagrangian and dynamic one-equation turbulence models available in OpenFOAM are assessed. Additional turbulence models (dynamic Smagorinsky and WALE), implemented by OpenFOAM community members, are adapted for incompressible flows and tested as well. In addition to this, the previously unavailable sigma-model was implemented in this work. Simulations without using any turbulence models, ie. under-resolved DNS (UDNS) simulations, were performed for comparison. Very good results were obtained in all cases with variations among the individual models. The no-model simulations performed surprisingly well and occasionally better than some of the models.

The parallel performance of the solver is tested on a computational cluster and the experiences gained during the course of this work are summarised as recommendations for future use.

Contents

Nomenclature

Symbol	Description
A	Area
c	Heat capacity
C	Generic coefficient
CFL	Courant-Friedrichs-Lewy stability number
D	Diameter
e	Reference length
E	Efficiency
f, F	Force or frequency
h, H	Reference length
I	Momentum flux ratio or invariant
IQ	Quality index
k	Thermal conductivity or kinetic energy
l	Reference length
L	Operator in subgrid models
M	Blowing Rate, operator in subgrid models or modelling amount
Nu	Nusselt number
OP	Generic operator
p	Pressure
phi	Face flux
Pr	Prandtl number
q	Heat flux
R	Generic ratio
Re	Reynolds number
S	Rate of strain tensor or parallel speedup

St	Strouhal number
t	Time
T	Temperature or elapsed time
u	Velocity
x, y, z	Spatial coordinates

Greeks

α	Angle or thermal diffusivity
β	Pressure drop per unit length
γ	Temperature source
δ	Kronecker delta symbol
Δ	Filter width
η	Adiabatic effectiveness
ν	Kinematic viscosity
ρ	Density
σ	Singular value
τ	Shear stress, stress tensor

Subscripts

∞, b	Bulk or farfield quantity
c	Coolant
cva	Cell volume weighted average
d	Drag
l	Lift
t	Turbulent
τ	Shear stress
v	Viscous
w	Wall

Superscripts

$'$	Fluctuation
$-$	Averaged or filtered quantity
$+$	Normalised quantity

List of Figures

List of Tables

1. Introduction

One of the major limitations when performing fluid dynamics simulation has been, and will always be, the available resources to do such calculations. In recent years the computational power of computers and the availability of large parallel clusters have drastically increased and with that, simulations have become more and more complex. This allows for more detailed studies of physical phenomena in challenging environments and geometries, but only if the software tools are enhanced along with the hardware improvements. The focus of this work lies on improving a previously developed solver (see [19]) to be used for the simulation of highly turbulent flows using a method known as large eddy simulation (LES).

Large eddy simulation has proved to perform very well in cases where turbulence is dominant, especially when the turbulent structures are anisotropic. The solver is therefore validated using cases where other, cheaper solution approaches for the Navier-Stokes equations, such as RANS (see below), have failed to produce accurate results. The simulation of cases involving heat transfer in incompressible flow will also be a part of this investigation, both for heat-transfer-enhancing geometries (turbulators) and for applications involving film cooling.

Since all these calculations require considerable computational power, the performance on a parallel cluster will also be investigated. Furthermore, quality aspects of computational grids and applicable numerical schemes will be treated as part of this work.

In the next chapters, the LES method itself and the models used in this study will be presented. The improvements implemented in the solver are explained in the chapters following the theory part.

2. Large Eddy Simulation

The behaviour of fluids can be described by the well-known mathematical model known as the Navier-Stokes equations. The original equations include formulations for the conservation of momentum, energy and mass, therefore leading to three momentum equations, one energy equation and one continuity equation. The form presented below is a simplification of these equations for incompressible flows and a constant viscosity. For a detailed derivation of the Navier-Stokes equations as well as the simplifications applied for incompressibility, many textbooks on fluid dynamics are available, eg. [12], [22] or [5].

$$\frac{\partial u_i}{\partial x_i} = 0 \tag{2.1}$$

$$\frac{\partial u_i}{\partial t} + \frac{\partial (u_i u_j)}{\partial x_j} = f_i - \frac{1}{\rho}\frac{\partial p}{\partial x_i} + \nu \frac{\partial^2 u_i}{\partial x_j \partial x_j} \tag{2.2}$$

Equation 2.1 shows the continuity equation and equation 2.2 shows the vectorial equation for momentum conservation. Here u_i is the velocity in direction i (x_i would be the corresponding spatial direction), f_i are body forces (eg. gravity), ρ is the fluid density, p the pressure and ν the laminar kinematic viscosity.

In turbulent flow simulations, the viscosity is often modified by turbulence models in the way that $\nu_{eff} = \nu + \nu_t$ where the laminar viscosity acting in the momentum equations will be replaced ν_{eff} (ν_t is a quantity calculated by the turbulence model and is called turbulent or eddy viscosity). In consequence the value of ν_{eff} will vary in space, and this requires that a part of the diffusion term of the momentum equations that was zero in the above equations now has to be retained (see [12] for a more detailed explanation). The momentum equation now takes the following form:

$$\frac{\partial u_i}{\partial t} + \frac{\partial (u_i u_j)}{\partial x_j} = f_i - \frac{1}{\rho}\frac{\partial p}{\partial x_i} + \frac{\partial}{\partial x_j}\left(\nu_{eff}\left(\frac{\partial u_i}{\partial x_j} + \frac{\partial u_j}{\partial x_i}\right)\right) \qquad (2.3)$$

One of the main difficulties when simulating flows is the presence of turbulence as it appears in a very broad spectrum of time and length scales. A proper simulation where all scales are fully resolved is known as DNS (direct numerical simulation). The computional effort to perform such simulations scales approximately with the Reynolds number cubed (see eg. [12]) and is currently only applicable to academic test cases at comparatively low Reynolds numbers and simple geometries.

A lot of effort has been put into developing formulations that model the behaviour of turbulence, therefore reducing the spatial and temporal resolution required to obtain a solution. The above equations are valid for incompressible flows with a spatially varying viscosity and form the basis for further simplifications of the Navier-Stokes Equations that allow the modelling of turbulence. One of these simplifications is the decomposition of the solution variables into an average and a fluctuating quantity, a procedure first described by Reynolds [36] leading to the Reynolds-Averaged Navier-Stokes (RANS) equations, shown below in equations 2.4 and 2.5. Their derivation and further explanations are omitted here.

$$\frac{\partial \bar{u}_i}{\partial x_i} = 0 \qquad (2.4)$$

$$\frac{\partial \bar{u}_i}{\partial t} + \frac{\partial (\bar{u}_i \bar{u}_j)}{\partial x_j} = \bar{f}_i - \frac{1}{\rho}\frac{\partial p}{\partial x_i} + \frac{\partial}{\partial x_j}\left(\nu\left(\frac{\partial \bar{u}_i}{\partial x_j} + \frac{\partial \bar{u}_j}{\partial x_i}\right)\right) - \frac{\partial \left(\overline{u_i' u_j'}\right)}{\partial x_j} \qquad (2.5)$$

Here, the additional term $\overline{u_i' u_j'}$, called the Reynolds stress tensor, needs to be modeled. This procedure has found widespread acceptance, especially in the industry, as it allows the calculation of stationary and simplified (eg. symmetric) solutions of the Navier-Stokes equations at greatly reduced computational costs compared to DNS. The main drawback of this method is that the fluctuating quantities of the solution variables are entirely calculated by turbulence models. Simulations where turbulent effects are dominant can often not be accurately represented by RANS turbulence models. Or it can be the case that a model performs very well in some cases but fails in others.

Another approach of dealing with turbulence is the so-called Large Eddy Simulation (LES), which is the main topic of this work. In contrast to the RANS method, the Navier-Stokes equations are not averaged, but filtered. A generic filter function (shown here in one-dimensional notation)

$$\bar{u}_i(x) = \int G(x, x') u_i(x') dx' \qquad (2.6)$$

is applied to the solution variables, where $G(x, x')$ is the filter kernel (see [12]). The filtered Navier-Stokes equation then take the following form:

$$\frac{\partial \bar{u}_i}{\partial x_i} = 0 \qquad (2.7)$$

$$\frac{\partial \bar{u}_i}{\partial t} + \frac{\partial(\bar{u}_i \bar{u}_j)}{\partial x_j} = \bar{f}_i - \frac{1}{\rho}\frac{\partial p}{\partial x_i} + \frac{\partial}{\partial x_j}\left(\nu\left(\frac{\partial \bar{u}_i}{\partial x_j} + \frac{\partial \bar{u}_j}{\partial x_i}\right)\right) - \frac{\partial \tau_{ij}^r}{\partial x_j} \qquad (2.8)$$

Here $\bar{u}$ is the filtered velocity, ie. the new solution variable. It can be seen that the continuity equation, due to its linearity, does not change. Similar however, to the Reynolds stresses in the RANS equations above, an additional term, the residual stress tensor τ_{ij}^r, can be found on the right hand side. This term appears due to the fact that when filtering the equations, the nonlinear convection term is actually

$$\frac{\partial(\overline{u_i u_j})}{\partial x_j}. \qquad (2.9)$$

Since term $\overline{u_i u_j}$ is not easily calculated, Leonard [20] suggested the split into $\overline{u_i u_j} = \tau_{ij}^r + \bar{u}_i \bar{u}_j$, therefore resulting in the final filtered forms given above in equations 2.7 and 2.8. The residual stress tensor τ_{ij}^r is unknown and needs to be modeled by so-called subgrid or LES turbulence models.

3. Subgrid Models

The following sections describe the individual models used in the context of this work and how they approximate these subgrid stresses. First and foremost, all models shown here are based on the eddy-viscosity assumption, ie. it is assumed that the effects of the subgrid stresses cause increased transport and dissipation and can therefore be approximated (Boussinesq-approximation, see [12]) by increasing the laminar viscosity by a turbulent counterpart. The following approximation is used:

$$\tau_{ij}^r - \frac{1}{3}\tau_{kk}^r \delta_{ij} = \nu_t \left(\frac{\partial \bar{u}_i}{\partial x_j} + \frac{\partial \bar{u}_j}{\partial x_i} \right) = 2\nu_t \bar{S}_{ij} \tag{3.1}$$

Here the left hand terms form the deviatoric part of the subgrid stress tensor τ_{ij}^r, ν_t is the turbulent viscosity and $\bar{S}_{ij}$ is once again the strain rate tensor of the resolved field. The task of the turbulence model is then to approximate the value of this turbulent viscosity or eddy viscosity.

The models presented below all have the following form when calculating the turbulent viscosity:

$$\nu_t = (C\Delta)^2 OP \tag{3.2}$$

where C is a model constant (or a combination of constants), Δ is the filter width and OP is a generic operator.

All of the models can be classified into different groups. The first distinction can be made based on how the constant C is treated. If the constant is an actual constant and remains the same over the course of the simulation then the models are called static models. In contrast to this, dynamic models locally calculate the constant, usually by re-filtering the resolved field using a filter width of 2Δ and comparing the differently resolved fields. A detailed explanation of this technique is omitted here, but can be found in eg. [8]. These new values for C can then be locally or globally averaged or clipped to avoid unphysical results causing instabilities (such as negative effective viscosities).

The models can be further distinguished by looking at the operator OP. Algebraic models simply solve an algebraic equation based on known quantities to calculate ν_t. N-equation models on the other hand solve additional transport equations, eg. for the turbulent kinetic energy, inside the operator. The filter width mentioned above Δ is a measure for the grid size and in most cases calculated as the cube root of the cell volume. All models presented below are in incompressible form therefore the density was generally dropped.

3.1 Smagorinsky Model

The Smagorinsky model was among the first models created for the purpose of calculating subgrid eddy viscosities, created in 1963 by Joseph Smagorinsky [41]. It is the most well-known and widely-used model and still applied today. It belongs to the category of static models where the C is an actual constant. Very good results can be obtained using this model when treating external detached flows and uniform grid turbulence.

The operator OP is simply defined as:

$$OP = |\bar{S}| \tag{3.3}$$

where $|\bar{S}| = (2\bar{S}_{ij}\bar{S}_{ij})^{\frac{1}{2}}$.

The main issue with the Smagorinsky model is that the constant C is not a universal constant but highly problem-dependent and needs to be calibrated to the case. This is especially problematic in the near-wall regions where the contribution of the turbulent viscosity should tend to zero. In these regions the Smagorinsky model drastically overestimates the turbulent viscosity.

One way around this problem is to apply a damping function to the constant C such as one suggested by van Driest [10] (originally used for RANS models):

$$C = C_0 \left(1 - e^{-y^+/A^+}\right)^2 \tag{3.4}$$

$$y^+ = \frac{\Delta y u_\tau}{\nu} \tag{3.5}$$

$$u_\tau = \sqrt{\tau_w} \tag{3.6}$$

Here C_0 is the desired coefficient in the main flow, y^+ is the dimensionless wall distance and u_τ is the friction velocity defined as the square root of the wall shear stress τ_w. The static, algebraic Smagorinsky model is provided as a built-in model in OpenFOAM.

3.2 Dynamic Smagorinsky Model

The dynamic version of the Smagorinsky model uses the same operator OP as above, but the coefficient C is calculated dynamically in each cell by re-filtering the resolved velocity field and comparing the results. More on the exact procedure can be found in [8].

The coefficient is given by

$$C = \frac{1}{2} \frac{L_{ik} M_{ik}}{M_{ik} M_{ik}} \tag{3.7}$$

where L_{ik} is equal to $-\overline{\bar{u}_i \bar{u}_k} + \bar{\bar{u}}_i \bar{\bar{u}}_k$ and M_{ik} is given as $\left(\frac{\bar{\Delta}}{\Delta}\right)^2 \left|\bar{\bar{S}}\right| \bar{\bar{S}}_{ik} - \overline{|\bar{S}| \bar{S}_{ik}}$. The double overline here indicates re-filtering using the test filter width $\bar{\Delta}$.

The value for C can reach very large negative values and in order to stabilise the solution the values are usually averaged, either over the domain volume or planes parallel to the flow. In complex geometries the plane averaging is not very meaningful and the averaging over the entire domain only properly works for homogenous turbulence. Another method is to average the values in a locally defined volume around the cell. The implementation used in this work does not use any averaging but clips the value for ν_t at $-\nu$, therefore producing zero viscosity at minimum.

9

This model was implemented by Alberto Passalacqua, an OpenFOAM community member [33], for compressible flow and was adapted here to an incompressible version.

3.3 Dynamic Lagrangian Model

This dynamic model was developed by Meneveau et al. [23] to improve the dynamic Smagorinsky models for flows in complex geometries. The averaging procedure of the dynamic Smagorinsky model was considered unsuitable for non-homogenous turbulence as the methods described in the previous section all require either homogenous directions (planes) or an arbitrary averaging volume. The idea was then to use lagrangian averaging along the streamlines of the flow, therefore being able to incorporate a fluid particles past trajectory. A short description of the method can be found in [44].

The operator OP is again the same as in the Smagorinsky model, but the coefficient C is calculated by solving two additional transport equations (not given here). The quantities calculated there have units of $[m^4/s^4]$ and require boundary and initial conditions just like other solution variables. Therein lies the difficulty when applying this turbulence model to non-periodic flow domains: the estimation of the inlet conditions for these variables is relatively arbitrary.

3.4 Dynamic One-Equation Model

The dynamic one-equation model in OpenFOAM solves a transport equation for the subgrid kinetic energy. The Smagorinsky models assumed that a local equilibrium exists between the transferred energy from the subgrid scales and the dissipated kinetic energy. If this is not the case, solving an additional transport equation for the subgrid kinetic energy can improve the results. Unfortunately the exact model used in OpenFOAM is not referenced and due to the large number of one-equation models it is unclear which one was actually implemented. For this reason, the equations shown here and some of the explanations are taken from the Fluent manual [2], which describes the model developed by Kim and Menon [17], as well as the OpenFOAM source code itself.

The operator OP for the calculation of the turbulent viscosity ν_t is given by

$$OP = \sqrt{k} \tag{3.8}$$

Here k is the subgrid kinetic energy defined as:

$$k = \frac{1}{2}\left(\overline{u_k^2} - \bar{u}_k^2\right) \tag{3.9}$$

where u is the resolved velocity and the overbar denotes filtering. The constant C is calculated using a dynamic procedure. The transport equation solved to evaluate k is given as follows:

$$\frac{\partial k}{\partial t} + \frac{\partial \bar{u}_j k}{\partial x_j} = -\tau_{ij}\frac{\partial \bar{u}_j}{\partial x_j} - C_\epsilon \frac{k^{\left(\frac{3}{2}\right)}}{\Delta} + \frac{\partial}{\partial x_j}\left(\nu_t \frac{\partial k}{\partial x_j}\right) \tag{3.10}$$

The additional constant C_ϵ contained in the k-equation is also calculated dynamically in OpenFOAM. Just as the dynamic Lagrangian model, the one-equation model is available built-in to the standard distribution of OpenFOAM.

3.5   WALE Model

The wall-adapting local eddy-viscosity (WALE) model was developed by Nicoud and Ducros [29] as a static, algebraic subgrid model that is able to reproduce the correct near-wall behaviour without the use of artificial damping functions (such as the van Driest damping described in the previous sections).

The model constant C is considered a true constant and a value of 0.5 is mostly used (the original paper suggests a value of C calculated as $\sqrt{10.6C_s}$ where C_s is the Smagorinsky constant).

The operator OP is based on the traceless symmetric part of the square of the velocity gradient tensor S_{ij}^d. The velocity gradient tensor is given as

$$\bar{g}_{ij} = \frac{\partial \bar{u}_i}{\partial x_j} \tag{3.11}$$

The overbar indicates that the resolved velocity is used for the gradient construction. The quantity S_{ij}^d mentioned above is then calculated as

$$S_{ij}^d = \frac{1}{2}\left(\bar{g}_{ij}^2 + \bar{g}_{ji}^2\right) - \frac{1}{3}\delta_{ij}\bar{g}_{kk}^2 \tag{3.12}$$

The operator OP of the WALE model is then constructed as

$$OP = \frac{\left(S_{ij}^d S_{ij}^d\right)^{3/2}}{\left(\bar{S}_{ij}\bar{S}_{ij}\right)^{5/2} + \left(S_{ij}^d S_{ij}^d\right)^{5/4}} \qquad (3.13)$$

where S_{ij}^d is given above and $\bar{S}_{ij}$ is the rate-of-strain tensor.

This model is widely used in commercial software and known for its good performance in transitional turbulent flows (an example is given in the original paper [29]). The OpenFOAM formulation of the operator was given by Cosimo Bianchini, an OpenFOAM community member, [4] and was completed here to a fully functional turbulence model.

3.6 Sigma Model

The sigma model was originally developed by Nicoud et al. [30] and is not available in OpenFOAM. It was implemented during the course of this work. The model is the spiritual successor of the previously described WALE model and has the same positive properties with some new ones added. The list of desirable properties for the operator OP that were used in the derivation in [30] is the following:

1. OP needs to be a positive quantity which involves only locally defined velocity gradients.

2. OP needs to follow cubic behaviour near solid walls (ie. tend cubically to zero).

3. OP needs to be zero for two-component or two-dimensional flows

4. OP needs to be zero for axisymmetric or isotropic expansion/contraction

The full derivation is omitted here, but as a final result the resulting operator was chosen to be

$$OP = \frac{\sigma_3(\sigma_1 - \sigma_2)(\sigma_2 - \sigma_3)}{\sigma_1^2} \qquad (3.14)$$

The constant C used in this model is considered to be a true constant and the suggested value for it is 1.5.

In equation 3.14 above, the σ-values are the singular values of the velocity gradient tensor. They are ordered in such a way that $\sigma_1 \geq \sigma_2 \geq \sigma_3 \geq 0$. This definition ensures that the first property mentioned above is met in the sense that OP will always be positive. This has the effect that the turbulent viscosity can never be negative and can therefore never lead to unphysical solutions and instabilities.

The second property is fulfilled as well since the second invariant of the velocity gradient tensor is quadratic in y, the wall normal direction. This fact is simply stated here without further explanation. A full discussion on the derivation of the σ values can be found in the original paper [30] and the numerical calculation of the values are given below.

The third property is given, as in the case of two-dimensional flows the smallest singular value σ_3 will always be zero. The fourth property is fulfilled as well, but is not as obvious as the previous one.

The implementation of this turbulence model is straightforward with the only difficulty being the calculation of the singular values of the velocity gradient tensor. OpenFOAM already offers a method to calculate the eigenvalues of tensors, from which the singular values can be obtained by taking the square root. During this work this built-in method has however failed several times causing floating point exceptions and stopping the simulation from running. This is most likely due to incomplete boundedness checks inside the algorithm. Therefore the following method, also suggested by [30], was implemented and boundedness of certain critical operations was enforced.

First, the velocity gradient tensor g is constructed for every cell. Then the invariants $I_1 = tr(g)$, $I_2 = \frac{1}{2}\left(tr(g)^2 - tr(g^2)\right)$ and $I_3 = det(g)$ are calculated, where tr denotes the trace and det the determinant of the tensor. Then the angles $\alpha_1 = \frac{I_1^2}{9} - \frac{I_2}{3}$, $\alpha_2 = \frac{I_1^3}{27} - \frac{I_1 I_2}{6} + \frac{I_3}{2}$ and $\alpha_3 = \frac{1}{3}arccos\frac{\alpha_2}{\alpha_1^{3/2}}$ are built. During this step it was ensured that the value of α_1 was positive and non-zero and that the argument of $arccos$ remained bounded between -1 and 1.

From that the singular values can be computed as follows:

$$\sigma_1 = \sqrt{\left(\frac{I_1}{3} + 2\sqrt{\alpha_1}cos\alpha_3\right)} \tag{3.15}$$

$$\sigma_2 = \sqrt{\left(\frac{I_1}{3} - 2\sqrt{\alpha_1}cos\left(\frac{\pi}{3} + \alpha_3\right)\right)} \tag{3.16}$$

$$\sigma_3 = \sqrt{\left(\frac{I_1}{3} - 2\sqrt{\alpha_1}cos\left(\frac{\pi}{3} - \alpha_3\right)\right)} \tag{3.17}$$

Here the arguments of the square root were forced to be non-negative but allowed to have a zero value.

4. Solver

4.1 Basics

In the previous work [19] a new solver for time-dependent incompressible flows was developed. It uses an explicit, third-order accurate time integration method based on the third-order Runge-Kutta scheme. It also offers the possibility of treating the viscous terms implicitly by applying a hybrid time integration scheme consisting of the Runge-Kutta and the Crank-Nicolson method (see [14]). It was shown that the solver is able to reach the predicted orders of accuracy by performing an order-verification study using the method of manufactured solution [38]. The full details of the numerical schemes as well as the verification procedure and results can be found in [19].

At the time the solver was limited in the sense that it did not incorporate any kind of turbulence modelling, restricting its use for practical flows. For this work the solver was considerably improved by adding a new procedure for pressure-velocity coupling as well as including the support for turbulent flows. In order to allow for the simulation of periodic geometries (ie. channel flows, ribs, bumps etc.) a source term for the momentum equations was included. An energy equation was also added for simulations involving the transport of a passive temperature scalar (eg. incompressible film cooling). Again, a source term is added to this equation to allow periodic simulations. The following sections provide the details of these additions.

4.2 Improvements

4.2.1. Pressure-Velocity Coupling

The new solver adds a selectable option to treat the pressure-velocity coupling in the manner proposed by Rhie and Chow [37].

The algorithm used at every substep in the previous version can be summarised as follows (note that in OpenFOAM the convection term is split into the velocity variable u and a flux field phi, see eg. [15]):

1. Setup momentum equation, collecting all the terms (convection, diffusion etc.) multiplied by the Runge-Kutta weights. Since the solver is explicit, everything but the temporal term goes to the right hand side RHS.

2. Create a vector field from the timestep, the right hand side and the current solution: $HbyA = \frac{1}{A_p}RHS + U$. A_p is the diagonal coefficient of the matrix and in case of an explicit solver simply equal to δt^{-1}.

3. Interpolate this vector field onto the cell faces and calculate the dot product with the face normal S_f to create a flux field $\overline{phiHbyA}$.

4. Solve pressure equation $lap(\frac{1}{A_p}p) = div(\overline{phiHbyA})$, where lap is the discrete laplacian operator and div the discrete divergence.

5. Update the flux field phi with the interpolated flux field from above and subtract the face flux field from the pressure matrix.

6. Update the existing velocity field by adding $\frac{1}{A_p}(RHS - grad(p))$, where $grad(p)$ is the discrete gradient of the new pressure field.

7. Update boundary conditions and proceed to the next substep.

The new algorithm uses the same Runge-Kutta framework as the old one, but the steps performed in the flux calculation are different. The algorithm is formulated in such a way that it solves for a pressure correction as opposed to the actual pressure as before. The steps are as follows:

1. Setup momentum equation. This step is identical to the one before.

2. The diagonal coefficient $\frac{1}{A_p}$ is interpolated to the faces to create $\overline{\frac{1}{A_p}}$.

3. A unit-length vector field e_d, pointing from the owner cell center to the neighbour cell center, is created.

4. The discrete gradient of the current pressure field is interpolated to the cell faces to create $\overline{grad(p)}$.

5. A new pressure gradient at the face is created: $gradp_f = \overline{grad(p)} - \overline{grad(p)}_{proj} + gradp_{sn}$. Here $\overline{grad(p)}_{proj}$ is the interpolated gradient projected onto e_d and $gradp_{sn}$ is the uncorrected (central differencing) gradient normal to the cell face.

6. The current velocity field is interpolated onto the faces $(\overline{u})$ and the flux field is then: $phi = \overline{u} \cdot S_f - \left(\left(\frac{1}{A_p}\left(gradp_f - \overline{grad(p)}\right)\right) \cdot S_f\right) + (phi^0 - \overline{u}^0)$. Here S_f is the cell face normal with length of the face area and phi^0 and $\overline{u}^0$ are values from previous substeps. This last term is only present in the second and third substep.

7. Solve pressure correction equation $lap(\overline{\frac{1}{A_p}}p_{corr}) = div(phi)$.

8. Update phi with the face flux field from the pressure correction equation.

9. Update pressure with pressure correction and update boundary conditions.

10. Update velocity with the gradient of the pressure correction.

The order verification performed in [19] was repeated using the new method and it can be shown that the method performs very well and achieves the projected third-order accuracy in time. Figure 4.2.1 shows the evolution of the error with decreasing time steps. The error plateau seen at lower timesteps is due to the spatial discretisation.

The following figures show the improvement in reducing pressure oscillations when using the new formulation compared to the old one. The figures were taken during the initial stages of the flow around a sharp edge. The case on the left in figure 4.2.2 uses the old formulation and the one on the right uses the new one.

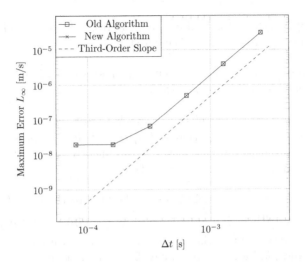

Figure 4.2.1: Temporal Discretisation Error

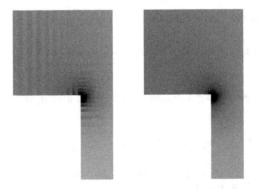

Figure 4.2.2: Reduction of Pressure Oscillations

4.2.2. Turbulence Models

General support for turbulence models was added in this step. This means that the solver works with any of the selectable turbulence models provided in OpenFOAM. The solver updates the turbulence fields between every substep of the Runge-Kutta scheme. This has one drawback: Since the

18

third-order time scheme is hardcoded in the solver, non-algebraic turbulence models can only be solved in second-order temporal accuracy. Only algebraic models can fully utilise the temporal accuracy of the solver. The error in the n-equation turbulence models will however be lower than when used in the default OpenFOAM solvers such as pisoFoam. This is due to the fact that the turbulence equations are solved multiple times in between every full timestep. It will be shown in the validation section (section 5) that n-equation models do not perform any worse compared to algebraic models. They do, however, require more computational effort.

4.2.3. Momentum Source Term

When simulating flows through channels and ducts periodic boundary conditions are usually applied in order to reduce the size of the computational domain. The periodicity condition for the velocity is easily imposed by simply treating the first and last cells in the domain as if they were direct neighbours. Periodic velocity is also physically valid in incompressible cases since the mass flow through the domain is always conserved. The pressure however drops from inlet to outlet due to wall friction and other energy losses and cannot simply be imposed on both sides of the periodic interface. To circumvent this problem a source term is added to the momentum equation that compensates for the pressure losses and corrects the velocity accordingly. Acharya [1] describes this procedure. The pressure is divided into a periodic part and a correction term:

$$p(x, y, z) = -\beta x + p_p(x, y, z) \tag{4.1}$$

Here $p_p(x, y, z)$ is the periodic part of the pressure and β is the channel pressure drop per unit length, ie. a pressure gradient. Acharya calculated the value of β using the current and the desired flow rate and applied over-relaxation factors to increase convergence for stationary simulations. The method used in this work is slightly different and common practice in fluid solvers.

First, the following term

$$\frac{\bar{u}}{|\bar{u}|} gp \tag{4.2}$$

is added to the right hand side of the momentum equation, where $\bar{u}$ is the desired bulk flow velocity and gp is the artificial pressure gradient in flow direction. The latter can usually be set to zero for the first iteration.

After solving the momentum and pressure equations, the cell volume weighted average of the velocity field in direction of $\bar{u}$ is calculated:

$$|\bar{u}|^* = (u \cdot \frac{\bar{u}}{|\bar{u}|})_{cva} \qquad (4.3)$$

Next, the difference of the current flow field and the target bulk velocity is calculated:

$$gp^+ = \frac{|\bar{u}| - |\bar{u}|^*}{(\frac{1}{A_p})_{cva}} \qquad (4.4)$$

where A_p is the diagonal coefficient of the momentum matrix and the subscript $_{cva}$ again denotes the cell volume weighted average.

The velocity field is then corrected using the pressure gradient corrector gp^+

$$u = u + \frac{\bar{u}}{|\bar{u}|} \frac{1}{A_p} gp^+ \qquad (4.5)$$

and the pressure gradient used in the momentum source term is also updated.

$$gp = gp + gp^+ \qquad (4.6)$$

These corrections are performed after every substep and therefore do not influence the temporal accuracy of the solver.

4.2.4. Energy Equation and Source Term

In order to be able to simulate flows including heat transfer, an additional energy equation was added to the solver. This is not an energy equation as it appears in the fully compressible Navier-Stokes equations but a simple scalar transport equation for the temperature. The temperature is transported as a passive scalar, meaning that it has no influence on the flow field. The procedure followed here is again based on the work by Acharya [1]. The equation is given below:

$$\frac{\partial T}{\partial t} + \frac{\partial (u_i T)}{\partial x_i} - \frac{\partial}{\partial x_i}\left(\alpha_{eff} \frac{\partial T}{\partial x_i}\right) = 0 \qquad (4.7)$$

Here T is the temperature, u is again the velocity and α_{eff} is the effective thermal diffusivity. This diffusivity is given by:

$$\alpha_{eff} = \frac{\nu}{Pr} + \frac{\nu_t}{Pr_t} \qquad (4.8)$$

where ν is the kinematic laminar viscosity and ν_t is its turbulent counterpart. Pr and Pr_t are the respective laminar and turbulent Prandtl numbers which have to be given as simulation parameters.

When simulating periodic flows with heat transfer, a source term (or sink term depending on the conditions) has to be added to equation 4.7 above. This source term can be derived from an energy balance (again see [1]) and has the following form:

$$u \cdot \frac{\bar{u}}{|\bar{u}|} \gamma \qquad (4.9)$$

where

$$\gamma = \frac{q}{\rho_{ref} c_{p,ref} H_{ref} |\bar{u}|} \qquad (4.10)$$

Here q is the total heat flux in or out of the domain, ρ_{ref}, $c_{p,ref}$ and H_{ref} are reference quantities for the density, the specific heat capacity and the length respectively. These quantities need to be defined when running a simulation. The reference length H_{ref} can be calculated as the domain volume divided by the heated surface.

5. Validation

5.1 Overview

The solver along with the previously described turbulence models were validated by performing a series of different test cases. These cases were selected in such a manner that different physical phenomena can be investigated. They are also cases where stationary RANS models usually give unsatisfactory results due to the inadequacy of the turbulence models.

The first case shown is a widely-used validation case for turbulence models and wall functions. It involves the simulation of fully wall-bounded, periodic flow between two flat parallel plates at a given Reynolds number. Velocity profiles, fluctuations and stresses can be investigated and compared to widely available DNS data. Next, the external flow around a square cylinder is investigated. Here the flow is fully detached and statistical values such as the Strouhal number can be compared to measurements. A mixture of attached and separated flow including heat transfer is shown in a ribbed channel. Nusselt numbers and reattachment lenghts are the primary focus in these simulations. The last case shown also involves energy transport and investigates the efficiency of film-cooling in inclined cooling holes at a low blowing rate. All the details and results on theses cases are shown in the following sections.

5.2 Periodic Channel Flow

5.2.1. Description

One of the oldest and simplest cases for numerical investigations is the flow between two parallel plates. It has been extensively researched and has the advantage that, due to its simple geometry, fully resolved DNS using spectral codes can be performed at reasonably high Reynolds numbers. One of the main contributors to the numerical data of such simulations is the work by Moser et al. [25]. The data from this publication is available to the public under [26] and provides the basis for all comparisons made in this work.

Figure 5.2.1 below shows a sketch of the geometry.

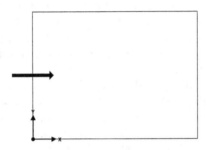

Figure 5.2.1: Channel Geometry Overview

Flows between parallel plates are usually characterised by a Reynolds number defined as:

$$Re_\tau = \frac{hu_\tau}{\nu}. \tag{5.1}$$

This Reynolds number is based on the friction velocity u_τ which, for incompressible flows, is given by $\sqrt{\tau_w}$, where τ_w is the wall shear stress. The length scale h is taken as the half-height of the channel and ν is the laminar kinematic viscosity. The specific case investigated in this work was performed at a friction Reynolds number of 395.

24

The original authors performed the simulation in a domain width edge lengths of $2\pi h \times 2h \times \pi h$. The size of this domain was chosen in such a manner that the largest occuring turbulent structures can be captured. The mesh resolution in the corresponding directions were $256 \times 193 \times 192$, resulting in a mesh size of approximately 9.5 million cells. The important criteria for such cases are the dimensionless distances in wall normal and wall parallel-directions (x is assumed to be the flow direction):

$$x^+ = \frac{\Delta x u_\tau}{\nu} \tag{5.2}$$

$$y^+ = \frac{\Delta y u_\tau}{\nu} \tag{5.3}$$

$$z^+ = \frac{\Delta z u_\tau}{\nu} \tag{5.4}$$

The Δ-values are the mesh sizes in the corresponding directions. The values for these dimensionless distances used in the original paper are given here for comparison with the values used in the present work:

x^+	y^+	z^+
10.0	0.0295 - 4.81	6.5

Table 5.2.1: Dimensionless Mesh Sizes of DNS Simulation

The values for x^+ and z^+ were taken from the original paper, whereas the values for y^+ were extracted from the numerical data. The lower value of y^+ corresponds to the first cell near the wall and the higher value represents the mesh size at the centerline of the channel.

5.2.2. Setup

The simulation domain size was reduced to $3.5 \times 2 \times 1.3\ h$ as suggested by other authors (eg. [30]). This reduction directly leads to decreased mesh size while not influencing the results too drastically.

As mentioned above, the case was run at a friction Reynolds number of 395, which leads to a target value for u_τ of 0.0079. A bulk velocity of 0.138 $[m/s]$ was enforced using the source term described in section 4.2.3. The value for the bulk velocity was obtained by integrating the velocity profile of the DNS data.

The following table 5.2.2 shows the boundary conditions for the case. The wall boundary condition implies a zero fixed velocity value and zero gradient value for all scalars except for the subgrid viscosity, which was also set to zero. The boundary condition for this turbulent viscosity is not critically important when the boundary layer is resolved, as the values at the wall are nearly zero when using the zero gradient condition.

Boundary	Type
inlet & outlet	translational periodic
front & back	translational periodic
top & bottom	no slip wall boundary

Table 5.2.2: Boundary Conditions for Periodic Channel Flow

5.2.3. Mesh

Two meshes were constructed for this case to observe the influence of the mesh resolution: a coarse mesh consisting of 51000 cells and a fine mesh with 414000 cells. The dimensionless mesh sizes for both meshes (based on the u_τ-value reported above) were as follows:

Mesh	x^+	y^+	z^+
Coarse	46.22	1.47 - 29.35	20.54
Fine	23.03	0.73 - 14.58	10.27

Table 5.2.3: Dimensionless Mesh Sizes of LES Simulation

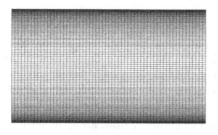

Figure 5.2.2: Channel Mesh Fine

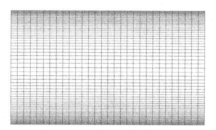

Figure 5.2.3: Channel Mesh Coarse

5.2.4. Results

The mean velocity and stress profiles were evaluated for all cases. Mean values of the velocity and the velocity fluctuations in all spatial directions were recorded during the simulation and averaged onto a single line. The the mean wall shear stress was recorded as well and averaged over all wall surfaces to provide a single value for the friction velocity u_τ.

The simulations were started from an artificial turbulence field in order to accelerate the transition to properly turbulent flow. Details of the initialisation procedure can be found in [8] If constant values are used for the initialisation, transition to turbulence is only triggered by numerical effects such as round-off errors and will take a very long time to occur. The simulation was allowed about 25 flow-through times to initialise and was then averaged over another 240. The time used here for the initialisation was comparatively high. In [8] it was shown that a lower initialisation time can be adequate as well.

All velocity values and their fluctuations, as well as the wall distance, were normalised using the value of u_τ in the following manner:

$$y^+ = \frac{\Delta y \bar{u}_\tau}{\nu} \tag{5.5}$$

$$u^+ = \frac{\bar{u}}{\bar{u}_\tau} \tag{5.6}$$

$$u'^+ = \frac{\bar{u}'}{\bar{u}_\tau} \tag{5.7}$$

$$v'^+ = \frac{\bar{v}'}{\bar{u}_\tau} \tag{5.8}$$

$$w'^+ = \frac{\bar{w}'}{\bar{u}_\tau} \tag{5.9}$$

$$\tau^+ = \frac{\bar{u}'\bar{v}'}{\bar{u}_\tau^2} = \frac{\tau}{\tau_w} \tag{5.10}$$

Here Δy is the wall normal distance from wall to cell center, ν is the molecular kinematic viscosity, $\bar{u}$ is the mean value of the x-component of the velocity, all primed values are averaged fluctuations and τ_w is the wall shear stress.

28

Coarse Mesh

First the results obtained using the coarse mesh are presented. Figures 5.2.4 to 5.2.7 show the mean dimensionless velocity profiles. In figures 5.2.8 to 5.2.11 the fluctuations and in 5.2.12 to 5.2.15 the shear stress profile can be seen.

The diagrams are ordered in such a fashion that the different models are put into meaningful groups. The values of the predicted friction velocity are provided in table 5.2.4. The results show that the dynamic models perform significantly better than any of the static models, even the wall-adapting WALE and sigma models. The static Smagorinsky model using the van Driest damping is not able to capture any of the meaningful quantities, especially in the case of the mean fluctuations. The best model for this grid is no model at all. The reason behind this is given in section 5.6.

Case	Friction Velocity u_τ $[m/s]$
Smagorinsky & VanDriest	0.00774
Dynamic Smagorinsky	0.00783
Dynamic Lagrangian	0.00776
Dynamic 1-Equation Model	0.00773
WALE	0.00741
sigma	0.00739
No Model	0.00781
DNS	0.00785

Table 5.2.4: Friction Velocity, Coarse Mesh

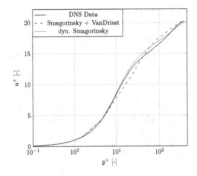

Figure 5.2.4: Mean Velocity Profiles, Smagorinsky Models

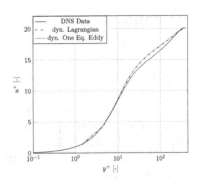

Figure 5.2.5: Mean Velocity Profiles, n-Equation Models

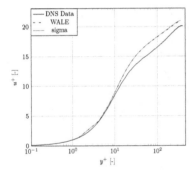

Figure 5.2.6: Mean Velocity Profiles, Static Models

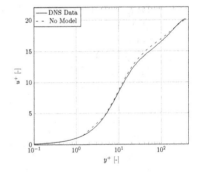

Figure 5.2.7: Mean Velocity Profiles, No Model

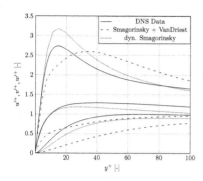

Figure 5.2.8: Mean Velocity Fluctuations, Smagorinsky Models

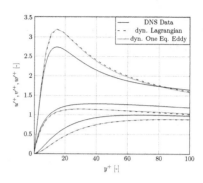

Figure 5.2.9: Mean Velocity Fluctuations, n-Equation Models

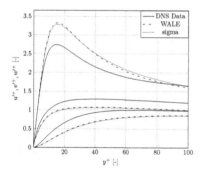

Figure 5.2.10: Mean Velocity Fluctuations, Static Models

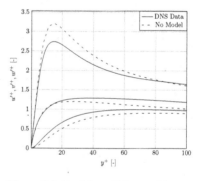

Figure 5.2.11: Mean Velocity Fluctuations, No Model

31

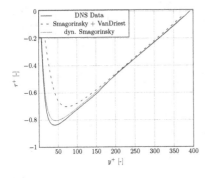

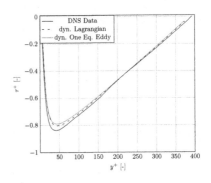

Figure 5.2.12: Mean Shear Stress, Smagorinsky Models

Figure 5.2.13: Mean Shear Stress, n-Equation Models

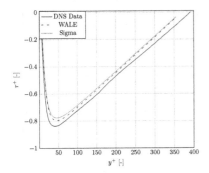

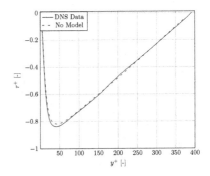

Figure 5.2.14: Mean Shear Stress, Static Models

Figure 5.2.15: Mean Shear Stress, No Model

32

Fine Mesh

The same evaluations have been performed on the fine mesh and the diagrams are grouped in the same way as before. All the results are much closer to the reference data provided by the DNS solution. The Smagorinsky model using the vanDriest damping function still shows some deficiencies which are due to the fact that the vanDriest damping is an artificial modifier to the subgrid viscosity and not physically meaningful. The velocity profiles of the dynamic Smagorinsky and the dynamic n-Equation models (figures 5.2.16 and 5.2.17) all agree very well with the data whilst the static models seem to slightly overestimate the subgrid viscosity, leading to higher velocities in the logarithmic region of the profile. The no-model simulation again shows surprisingly good results while very slightly underestimating the profile. This indicates simply that for a DNS simulation the grid is still too coarse.

The velocity fluctuations and shear stresses are better represented here compared with the results from the coarse grid. The agreement is very good close to the wall but quickly deteriorates further towards the channel centerline. This effect is observed in most papers doing such comparisons (eg. [8] [30]).

The values of the friction velocity on the fine mesh are shown below (table 5.2.5).

Case	Friction Velocity u_τ $[m/s]$
Smagorinsky & VanDriest	0.00787
Dynamic Smagorinsky	0.00793
Dynamic Lagrangian	0.00791
Dynamic 1-Equation Model	0.00791
WALE	0.00779
sigma	0.00776
No Model	0.00797
DNS	0.00785

Table 5.2.5: Friction Velocity, Fine Mesh

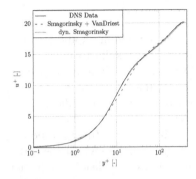

Figure 5.2.16: Mean Velocity Profiles, Smagorinsky Models

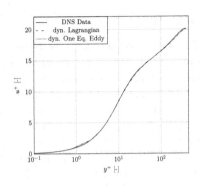

Figure 5.2.17: Mean Velocity Profiles, n-Equation Models

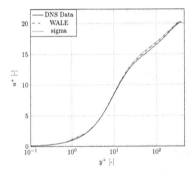

Figure 5.2.18: Mean Velocity Profiles, Static Models

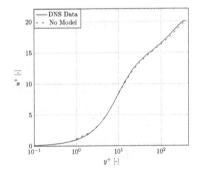

Figure 5.2.19: Mean Velocity Profiles, No Model

34

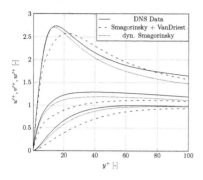

Figure 5.2.20: Mean Velocity Fluctuations, Smagorinsky Models

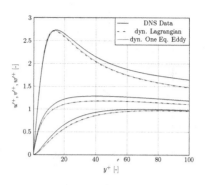

Figure 5.2.21: Mean Velocity Fluctuations, n-Equation Models

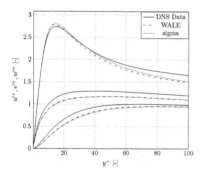

Figure 5.2.22: Mean Velocity Fluctuations, Static Models

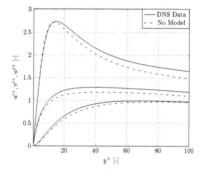

Figure 5.2.23: Mean Velocity Fluctuations, No Model

35

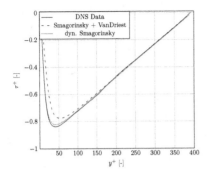

Figure 5.2.24: Mean Shear Stress, Smagorinsky Models

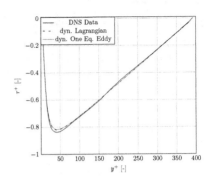

Figure 5.2.25: Mean Shear Stress, n-Equation Models

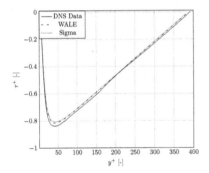

Figure 5.2.26: Mean Shear Stress, Static Models

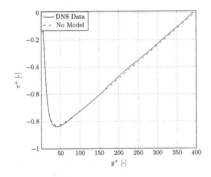

Figure 5.2.27: Mean Shear Stress, No Model

5.3   Square Cylinder

5.3.1. Description

The external flow around bluff bodies is the focus of many research investigations. The case shown here deals with a square cylinder of dimensions
$e \times e \times 4e$ (where $e = 0.04$ $[m]$) in free flow that was experimentally investigated by Lyn et al. [21],[7]. The cylinder is placed in a three-dimensional
channel of dimensions $0.82 \times 0.56 \times 0.16$ $[m]$. Figure 5.3.1 shows a sketch of
the simulation domain as well as the location of the origin of the coordinate
system used for all the evaluations below.

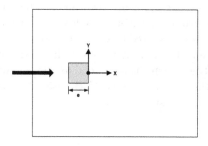

Figure 5.3.1: Channel Geometry Overview

The main interest here is to be able to accurately predict aerodynamic
properties of the body such as Strouhal number and drag coefficient. The
geometry is simple and offers well-defined separation points for the flow at
the sharp edges of the cylinder.

5.3.2. Setup

The following table 5.3.1 provides an overview of the boundary conditions
used. The inlet velocity was given by the original paper [21] and had a
value of 0.535 $[m/s]$. The kinematic viscosity ν was set to 1e-5 $[m^2/s]$
and the timestep used was 0.001 $[s]$. The simulation was run over a total
duration 50 $[s]$, ie. 50'000 timesteps.

Boundary	Type
inlet & outlet	fixed value/zero gradient
front & back	translational periodic
cylinder	no slip wall boundary
top & bottom	symmetry plane

Table 5.3.1: Square Cylinder Boundary Conditions

Note that the simulation is fully three-dimensional, 2D-simulations were tested but did not give any meaningful results. This is obvious, as in large eddy simulations, a large part of the turbulent spectrum is resolved, and turbulent structures are always three-dimensional. In fact, in the sigma model it is even the case that in locally two-dimensional or two-component flows, the contribution from the turbulent eddy viscosity is zero.

Constant initial values were used to start the case as, due to the enforced separation, the turbulent flow field develops rather quickly. A slightly distorted initial field for the velocity is beneficial as the separation can remain symmetrical around the x-axis for several thousand steps and only then develop into a proper vortex street.

5.3.3. Mesh

Due to the simplicity of the geometry, a fully orthogonal, hex-cell mesh was created. The resolutions in the three spatial coordinates were 160 × 198 × 32, resulting in an overall mesh size of 962'560 cells. A frontal view of the mesh can be seen in figure 5.3.2. The near wall resolution of the mesh was rather coarse, resulting in average values of approximately $x^+ = y^+ = 5$ and $z^+ = 25$. As the flow is mostly detached and the separation points are given by the geometry, increasing the resolution in wall normal direction will not significantly improve the results but will cause issues for the allowable timestep limit.

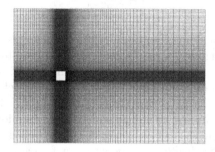

Figure 5.3.2: Square Cylinder Mesh

5.3.4. Results

In this case the Strouhal number St, the drag and lift coefficients C_d and C_l as well as the averaged velocity were evaluated. These quantities are defined as follows:

$$St = \frac{f l_{ref}}{u_\infty} \tag{5.11}$$

$$C_d = \frac{F_x}{\frac{1}{2} u_\infty^2 A_{ref}} \tag{5.12}$$

$$C_l = \frac{F_y}{\frac{1}{2} u_\infty^2 A_{ref}} \tag{5.13}$$

Here f is the shedding frequency of the flow, l_{ref} is a reference length (in our case the edge length of the cylinder, ie. 0.04 $[m]$) and u_∞ is the freestream velocity, set to 0.535 $[m/s]$. In the drag coefficients F is the force in the respective coordinate direction and A_{ref} is a reference area which here is equal to the cross-sectional area of 0.0064 $[m^2]$. The total forces F are evaluated by summing up the pressure and friction forces. These forces were used not only in the calculation of the drag coefficients but also in the determination of the shedding frequency f. A frequency analysis via FFT (Fast Fourier Transform) was performed on the signal of the forces in the y-direction. This is also the reason why the simulation was run for a rather long time as the resolution of the frequency spectrum depends on the total number of available samples.

The first 5'000 timesteps were not used in the evaluation in order to allow the flow to fully develop first.

In figures 5.3.3 to 5.3.6 below the results of the averaged centerline velocity in x-direction are shown and compared to the measurements by [7]. Compared to the original paper, the origin of the x-axis is shifted by 0.5h, ie. the origin is not in the center of the cylinder but just at the end where the wake begins.

As the original paper does not provide any information on the fluctuation of the drag coefficients and only provides mean values, simulation results by other authors are listed here as well. Ochoa [32] provides a helpful overview of many such comparisons.

This overview is listed in table 5.3.2 below.

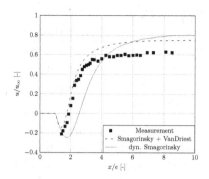

Figure 5.3.3: Averaged Velocity Profile, Smagorinsky Models

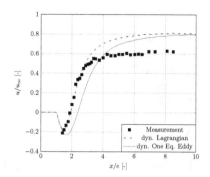

Figure 5.3.4: Averaged Velocity Profile, n-Equation Models

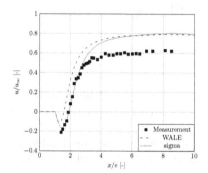

Figure 5.3.5: Averaged Velocity Profile, Static Models

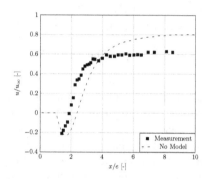

Figure 5.3.6: Averaged Velocity Profile, No Model

It can clearly be seen in figures 5.3.3 to 5.3.6 that none of the models are able to predict the final recovery value of the mean velocity. This is mostly due to the mesh resolution. The fact that constant boundary values were used, and therefore no inlet turbulence level was present, could also have influenced this result. Since the no-model simulation shows the same behaviour, the issue cannot be attributed to the models themselves. The dynamic lagrangian and the sigma model are the only ones that are able to capture the location and magnitude of the steep gradient accurately and agree very well with the measured values. As this region in the flow

Case	St	$\overline{C_l}$	C_l'	$\overline{C_d}$	C_d'
Smagorinsky & vanDriest	0.128	-0.017	1.442	2.344	0.220
Dynamic Smagorinsky	0.138	-0.010	0.521	1.769	0.062
Dynamic Lagrangian	0.139	-0.001	1.011	1.993	0.081
Dynamic One-Eq. Eddy	0.138	0.01	0.646	1.826	0.058
WALE	0.135	0.008	1.429	2.239	0.217
sigma	0.140	0.001	1.033	2.004	0.080
No Model	0.139	-0.008	0.615	1.794	0.059
Verstappen and Veldman [43]	0.133	0.005	1.45	2.09	0.178
Porquie et al. [35]	0.13	-0.02	1.01	2.2	0.14
Murakami et al. [27]	0.131	-0.05	1.39	2.05	0.12
Wang and Vanka [45]	0.13	0.04	1.29	2.03	0.18
Nozawa and Tamura [31]	0.131	0.009	1.39	2.62	0.23
Ochoa and Fueyo [32]	0.139	0.03	1.4	2.01	0.22
Exp.: Lyn et al. [7] [21]	0.132	-	-	2.1	-

Table 5.3.2: Comparison of Aerodynamic Data

is the most important for the aerodynamical drag around bluff bodies it comes as no surprise that these two models are also able to predict the drag coefficient well.

The Strouhal number was comparably close to the measured value in all of the cases and in the same range as obtained by Ochoa and Fueyo [32]. However, other authors have been able to reach values that are closer to the measurements.

In order to improve the simulation results, the mesh would have to be refined, especially in the wake region. It would also improve the prediction of the Strouhal number. This point is explained further in chapter 5.6 below.

5.4 Ribbed Channel

5.4.1. Description

In flows where heat transfer is important, eg. for cooling certain machine parts, high heat transfer coefficients are always desirable. One way to achieve such increased heat transfer coefficients is by adding turbulators. Turbulator is a generic term for any geometric modification of a surface that increases turbulence levels, thereby improving mixing and energy transfer into the core flow. The case investigated here involves such a geometry of a rectangular turbulator in a periodic channel. The overall domain size is $0.127 \times 0.061 \times 0.060325 \ [m]$ and height and width of the turbulator rib, which lies in the center of the bottom wall, is $e = 0.00635 \ [m]$. The case was originally investigated by [1] whose measurements served as the basis for comparisons in this work. A sketch of the geometry and the location of the coordinate system origin can be seen in figure 5.4.1.

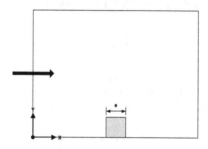

Figure 5.4.1: Ribbed Channel Geometry Overview

5.4.2. Setup

In this case, the temperature transport equation described in section 4.2.4 was used to simulate heat transfer. Both the momentum and energy source terms had to be activated in order to allow for periodic boundary conditions. Table 5.4.1 below shows the boundary conditions used in this case.

43

Boundary	Type
inlet & outlet	translational periodic
front & back	translational periodic
bottom wall & rib	no slip wall boundary
top wall	no slip wall boundary

Table 5.4.1: Ribbed Channel Boundary Conditions

Additionally, the bottom wall, excluding the rib, was a heated wall and required special boundary conditions for the temperature equation. One way to achieve a heating wall would be to simply set a fixed wall temperature value. The drawback of this method is that the total heat flux into the domain would depend on the flow field, ie. the γ term in equation 4.9 would have to be recalculated for every timestep. For this reason the total heat flux was fixed to a value of 280 $[W/m^2]$, therefore keeping γ a constant value. In order to achieve this, a fixed gradient boundary condition was imposed on the bottom wall, using a gradient value of 10415.7135. This value was calculated through the following relation (see [2]):

$$q = k_f \left(\frac{\partial T}{\partial n} \right)_{wall} \tag{5.14}$$

Here q is the wall heat flux, k_f is the thermal conductivity of the fluid and the last term is the wall normal gradient of the temperature we are looking for. The conductivity can be calculated as

$$k_f = \alpha \rho c_p \tag{5.15}$$

where α is the thermal diffusivity, ρ is the density and c_p is the specific heat capacity at constant pressure.

In our solver, α is calculated as

$$\alpha = \frac{\nu}{Pr} + \frac{\nu_t}{Pr_t} \tag{5.16}$$

where ν and ν_t are the laminar and turbulent viscosities and Pr and Pr_t the respective Prandtl numbers. In our case, ν_t, the subgrid viscosity, is set to zero at the wall and no wall functions are used. For this reason the turbulent contribution to the thermal diffusivity at the wall is zero and the gradient can be imposed as a constant. If this were not the case, the wall heat flux would have to be constantly updated to maintain the total required heat flux.

Care has to be taken that the wall gradient is calculated using the same quantities as used in the temperature equation source term. The bulk flow velocity was set to 3.6 $[m/s]$, leading to a Reynolds number of 28'341 (based on D_h). The rest of the quantities used are summarised in table 5.4.2 below.

Quantity	Value
ν	1.55e-5 $[m^2/s]$
ρ	1.208 $[kg/m^3]$
c_p	1005.2 $[J/(kgK)]$
H	0.06388 $[m]$
Pr	0.7 $[-]$
Pr_t	0.5 $[-]$
q	280 $[W/m^2]$
D_h	0.122 $[m]$

Table 5.4.2: Physical and Geometrical Data used in Ribbed Channel Case

The reference length H was calculated as the total domain volume divided by the heated surface area. The turbulent Prandtl number was set following a suggestion by Moin et al. [24] and twice the channel height was used for the hydraulic diameter, as is the case for parallel plates.

A timestep of 1e-5 $[s]$ was used and the simulation was run over the duration of 1 $[s]$ with the averaging procedure starting after 0.1 $[s]$. All quantities were initialised as constant values. Due to the presence of the rib and the resulting flow separation fully turbulent state is quickly reached and no addition of artificially turbulent fields was necessary.

5.4.3. Mesh

The mesh resolutions in the three coordinate directions were 136 × 104 × 33, amounting to a total cell number count of 493'152. The y^+ value was kept below 1 on the heated walls and varies over the rib. The values for x^+ and z^+ vary strongly due to the different mesh densities but never surpass a value of approximately $27 \times y^+$. Figure 5.4.2 gives an overview of the mesh.

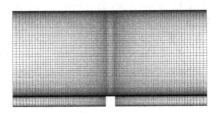

Figure 5.4.2: Ribbed Channel Mesh

5.4.4. Results

Nusselt Number Distribution

In order to assess the performance of the turbulator geometry, the Nusselt number Nu is evaluated. It is defined as the ratio of the convective and the conductive heat transfer, therefore providing insight on the effectivity of turbulator. It is defined as follows:

$$Nu = \frac{qD_h}{k(T_w - T_b)} \qquad (5.17)$$

Here q is the wall heat flux, D_h the hydraulic diameter, k the thermal conductivity, T_w the temperature at the wall and T_b is the bulk temperature. It is evident that when the properties and the heat flux are constant, a higher Nusselt number indicates a smaller temperature difference between the wall and the bulk flow, meaning better heat transport. The bulk temperature T_b is not a constant but depends on the simulation properties and the turbulence model used. It was evaluated by taking the mass flow average at the inlet periodic surface based on the time-averaged quantities.

The following figures 5.4.3 to 5.4.6 show the Nusselt number distribution over the position in flow direction. The position in x-direction was normalised using the height of the rib and the rib is located at a value between 9.5 and 10.5. The measurement values used for the comparison were taken from [1]. The wall temperatures evaluated were averaged over time and then averaged over the lateral direction to provide a single line data source.

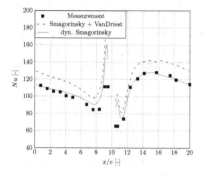

Figure 5.4.3: Nusselt Number Distribution, Smagorinsky Models

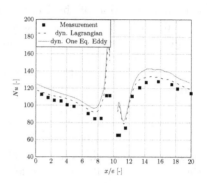

Figure 5.4.4: Nusselt Number Distribution, n-Equation Models

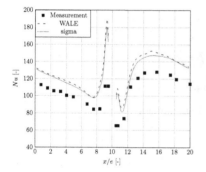

Figure 5.4.5: Nusselt Number Distribution, Static Models

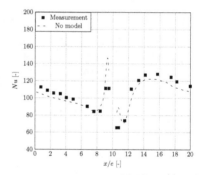

Figure 5.4.6: Nusselt Number Distribution, No Model

All of the models tested are able to capture the overall shape of the Nusselt number distribution. The dynamic Smagorinsky model agrees very well with the measurements whilst its overall profile is slightly compressed, underestimating the gradient before the rib. The static WALE and sigma models predict the profile very well but the results are offset compared to the measurements. This indicates a general over-estimation of the turbulent subgrid viscosity and therefore the effective thermal diffusivity. The same can be said for the dynamic one-equation model. According to [1] the maximum location of the Nusselt number after the rib lies at a position of 15.6 (with an measuring uncertainty of $\pm$ 0.4). This is predicted well by all models, however the exact location is difficult to determine in some cases (eg. Smagorinsky + van Driest) as the solution contains wiggles in this region. These could be evened out by averaging over a longer period of time.

Again the no model simulation shows very good results while slightly underestimating absolute values.

Just after the rib, in between locations 10.5 and approximately 11, a strong increase in the Nusselt number can be observed that is not present in the measured results. This is due to a local vortex in the vicinity of the rib which increases the energy transport. Figure 5.4.7 shows the streamlines of the averaged velocity in the wake region of the rib. It is possible that due to the way the measurements were taken ([1]) this local peak was smoothed out.

Figure 5.4.7: Averaged Streamlines

Reattachment Length

Also investigated in this case was the reattachment length behind the rib. For this reason, the wall shear stress was calculated at every time step and averaged along with the other quantities. The reattachment point was then evaluated by looking at the change of sign of the wall shear stress in flow direction. The reattachment location of separated flows are notoriously difficult to accurately represent when using Reynolds averaged (RANS) turbulence models.

Table 5.4.3 below shows the predicted reattachment lengths of all cases. It is evident that all models are well within the accuracy of the measurements and significantly better than RANS predictions conducted in an ongoing study at the University of Applied Sciences & Arts Lucerne. Due to the wide error margin of the measurements it is difficult to assess which of the models actually performed best.

Case	Reattachment length x/e $[-]$
Smagorinsky & VanDriest	6.015
Dynamic Smagorinsky	6.045
Dynamic Lagrangian	5.935
Dynamic 1-Equation Model	5.8
WALE	5.775
sigma	5.945
No Model	5.985
Measurement	6 ± 0.7

Table 5.4.3: Reattachment Lengths

5.5 Film Cooling

5.5.1. Description

The final case treated in this work deals with film cooling at low blowing rates. Film cooling is a practice often used in turbomachines, specifically gas turbines. The first stages of such turbines are exposed to very high temperature gases, often beyond the allowable limit of the turbine blade materials. For this reason, complex cooling geometries are built into the blade casings. One of the methods used in such situations is film cooling. Here gas is extracted from the machine at lower temperatures and blown into the critical parts of the machine. This cool gas then forms a protective layer around the geometry, thus avoiding the destruction of the materials. A quantity often used for classifying such cooling systems is the blowing rate, defined as:

$$M = \frac{\rho_c u_c}{\rho_\infty u_\infty} \tag{5.18}$$

where ρ_c and ρ_∞ are the densities and u_c and u_∞ are the velocities of the cooling flow and the main flow respectively. The values for the cooling flow part are evaluated inside the hole.

The prediction of such cooling layers is a difficult task, especially at low blowing rates where the main flow strongly interacts with the jet. Turbulence in such regions is highly anisotropic making it virtually impossible for isotropic RANS models to accurately predict the covered surface. RANS turbulence models are usually modified to account for the anisotropy, such as in [3]. An attempt was made in this work to employ a LES procedure to such a test case. Figure 5.5.1 below shows the geometry and the position of the coordinate system's origin.

5.5.2. Setup

The case investigated here follows the measurements conducted by Sinha et al. [40]. The case and geometry details can be found in the following table 5.5.1.

The first thing to note is that in this work an incompressible solver was used, therefore no density is available. The Mach numbers are low enough for this to be valid but it poses the question on how to replicate the flow conditions accurately. Johnson et al. [16] discuss the importance of

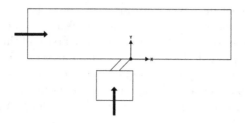

Figure 5.5.1: Film Cooling Geometry Overview

Quantity	Value
Hole Diameter	0.0127 $[m]$
Hole inclination	35 $[°]$
Diameter to Length Ratio	1.75 $[-]$
Channel Height	0.127 $[m]$
Bulk Flow Velocity	20 $[m/s]$
Cooling Flow Velocity	5 $[m/s]$
Blowing Rate	0.5 $[-]$
Bulk Flow Temperature	302 $[K]$
Cooling Flow Temperature	153 $[K]$
Bulk Flow Density	1 $[kg/m^3]$
Cooling Flow Density	2 $[kg/m^3]$

Table 5.5.1: Film Cooling Setup Parameters

different similarity criteria. This leads to the following possible quantities to keep constant:

1. The velocity ratio $R_u = \frac{u_c}{u_\infty}$

2. The blowing rate $M = \frac{\rho_c u_c}{\rho_\infty u_\infty}$

3. The ratio of momentum fluxes $I = \frac{M^2}{R_\rho}$, where R_ρ is the density ratio

In the first case, the velocities remain the same therefore no adaptation needs to be performed. In the second case, keeping the blowing rate constant, the velocity in the cooling hole needs to be doubled and when keeping the momentum fluxes constant, the cooling hole velocity has to be increased by a factor of $\sqrt{2}$. It is a priori not clear which of these methods best represents the compressible case. Here the first and third method were chosen as they represent the minimum and the maximum.

Only one subgrid model, the sigma model (see section 3.6), was used here as the computational effort for such a simulation is considerable.

The boundary conditions used are given in the following table 5.5.2:

Boundary	Type
inlet & outlet	fixed value / zero gradient
front & back	translational periodic
bottom wall & cooling hole wall	no slip wall boundary
top wall	symmetry plane

Table 5.5.2: Film Cooling Boundary Conditions

The inlet boundary condition was chosen to be a constant fixed value for the velocity, which is also used in RANS simulations of the present case. Using a periodic mapped boundary condition, simulating an infinite channel at the inlet, would have been the preferred method. Numerical instabilities prevented this, however. More on this topic can be found in chapter 9 on page 71.

The timestep used in this simulation was adapted automatically based on a fixed maximum CFL number of 1, which lead to time step sizes of around 2.3e-6. The simulation was started from constant fields and was run for 0.1 seconds, corresponding to about 40'000 timesteps, before the averaging procedure was started. The averaging was then performed over two flow-through times of the bulk flow.

5.5.3. Mesh

Three meshes were generated for this case but only one was used in the end. The coarsest mesh, consisting of only 266'728 cells, was in fact too coarse and did not deliver any meaningful results. The large variation in cell densities occasionally even lead to numerical instability. A finer and more even mesh was created using the same topology, resulting in a mesh size of just under 1.4 million cells. The finest mesh generated used a different topology and was smoothed to improve the overall orthogonality of the cells. This mesh consisted of about 2.5 million cells. Due to the very low timesteps required and the overall mesh size, the finest grid was not used and the results shown below were obtained using the medium mesh. Figures 5.5.2 and 5.5.3 show the coarse and the medium mesh respectively and on figure 5.5.4 a close-up of the fine mesh region around the cooling hole exit can be seen.

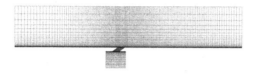

Figure 5.5.2: Film Cooling Mesh, Coarse, Regular

Figure 5.5.3: Film Cooling Mesh, Medium, Regular

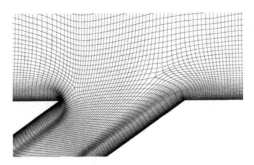

Figure 5.5.4: Cooling Hole Mesh, Fine, Smoothed

5.5.4. Results

The main interest when doing film cooling simulations is the surface coverage and the lateral spreading of the cool gas. In order to quantify this, the adiabatic effectiveness η is introduced, which is defined as follows:

$$\eta = \frac{\bar{T} - T_\infty}{T_c - T_\infty} \tag{5.19}$$

Here $\bar{T}$ is the temporally averaged temperature of the simulation and T_∞ and T_c are the bulk and cooling hole flow temperature respectively. The value of η ranges between 0 and 1, where 0 means no coverage and 1 indicates full coverage of the surface by the cool gas. For all diagrams below, the coordinate directions have been normalised by the hole diameter. Figures 5.5.5 and 5.5.6 below show the centerline effectiveness as well as the laterally averaged centerline effectiveness over the normalised streamwise direction.

The following two figures 5.5.7 and 5.5.8 show the lateral distribution of η at a normalised streamwise position of 6 and 15, respectively.

The results show that neither of the setups used accurately describes the centerline and laterally averaged effectiveness. Keeping R_u constant has lead to an overprediction of η, while keeping M constant leads to flow separation just after the cooling hole exit and therefore a locally reduced effectiveness. This leads to the conclusion that keeping the momentum flux ratio I constant would have been a better choice.

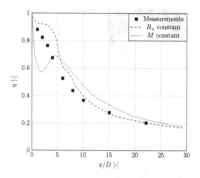

Figure 5.5.5: Centerline Adiabatic Effectiveness

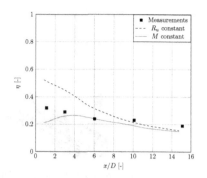

Figure 5.5.6: Laterally Averaged Adiabatic Effectiveness

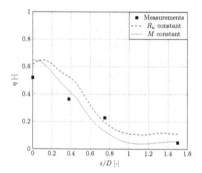

Figure 5.5.7: Adiabatic Effectiveness at x/D = 6

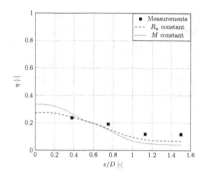

Figure 5.5.8: Adiabatic Effectiveness at x/D = 15

In figure 5.5.7, where the lateral distribution of η is shown, a dip in the effectiveness can be seen between $z/D = 0.6$ and 1.2. This dip is due to a horseshoe-vortex forming around the jet entering the main flow and transporting hot gas into the region past the cooling hole (see figure 5.5.9).

The effect can be seen in both cases investigated here and is therefore not due to the inlet velocity of the cooling gas. The reason for the strong effect of this vortex is most likely the lack of turbulent mixing, allowing the vortex to fully develop. This is a problem related to the boundary condi-

Figure 5.5.9: Horseshoe Vortex Around Cooling Hole Exit

tions used in the main flow, which were set to uniformly constant values. Due to the relatively short length of the domain inlet, flat-plate transition does not occur (which the sigma turbulence model is able to reproduce) and the flow arriving at the cooling hole is fully laminar. The reason why constant values were used at the inlet is given in section 9. In a future simulation, turbulent boundary conditions and a constant momentum flux ratio I will be applied and the results are expected to improve further.

5.6 Comments on the Results

5.6.1. No-Model Results

A few general comments on the above results need to be made. First we have the surprisingly good results of the no-model simulations. This is a phenomenon that has been observed by other authors as well (see eg. [6]) and is, after all, not that surprising. In this work we considered subgrid models that are all based on the eddy viscosity concept by Boussinesq (see eg. [12]). This means that all those models increase the effective viscosity of the fluid considered and therefore increase diffusion. An exception to this would be the localised dynamic Smagorinsky model where the turbulent viscosity can become negative and is clipped at the negative laminar viscosty (ie. eliminating the diffusion term fully). But even that model has an overall positive value of the subgrid viscosity and therefore increases diffusion. Numerical diffusion has exactly the same effect. We are using a diffusion-free central differencing scheme for the convection term, but since the turbulent flow field in large eddy simulations is hardly ever aligned with the computational grid, diffusion occurs nevertheless in finite volume methods. Looking at the results from our simulations we can therefore deduce that the numerical diffusion is more or less of the same order of magnitude as the contribution from the subgrid models. This also explains why all the dynamic models provide much better results than the static ones in the channel simulations: The dynamic models adapt their coefficients taking into account the numerical diffusion as well, since they perform an additional filtering of the resolved velocity field.

5.6.2. Grid Resolution

The second point to make is the lack of high-order spatial schemes in Open-FOAM. It has shown in this work that using only second-order schemes for large eddy simulations has several major drawbacks, all related through the grid resolution.

The first point to mention is the grid resolution required for accuracy: When using low-order spatial schemes, the grid resolution has to be very fine in order to adequately reduce the discretisation error. This directly leads to smaller time steps required to stay in the order of a CFL number of around one. This means that using low-order schemes increase calculation times significantly.

Secondly, the grid density required can reach a level where the influence of the subgrid models practically vanishes, therefore almost invalidating the use of any subgrid models. Again the example of the channel flow shall be mentioned here, especially in the case of the sigma model. The original paper [30] is able to obtain much better results with the model on a grid with a resolution that lies between the coarse and the fine version used here. This is mainly due to the fact that they were able to use fourth and fifth order schemes in their spatial discretisation (note however that other issues such as implementation and clipping will certainly also play a role here).

The last point to mention is that if a coarser grid is used, the spatial discretisation error might be at a level where the third order time integration scheme will not be able to fully deliver its potential. Figure 4.2.1 on page 18 shows the error evolution with respect to the time step obtained in the order verification study performed in [19]. The error value of the plateau on the left hand side is given by the spatial discretisation error, ie. the discretisation scheme and the grid resolution. If the grid is coarsened, the level of this plateau rises and may make the use of a (slower) third order time scheme pointless. It is possible that this is the reason why hardly any of the simulations performed in this work were able to accurately capture the Strouhal number of the square cylinder (reduction of the timestep did not improve the results in that respect).

6. Parallel Performance

Large eddy simulations are computationally expensive. Apart from a fine mesh resolution a large number of timesteps is required to get significant results. This is due to the fact that not only does the simulation have to reach a statistically steady state (eg. fully developed flow), it then needs to run long enough to provide meaningful average values. This is even more the case when transient statistics such as shedding frequencies are needed (eg. around a square cylinder, see section 5.3) as the resolution of the frequency spectrum depends on the total number of samples available. For these reasons parallel computing is a necessary tool in order to deal with large eddy simulations.

One of the most important qualities of an LES solver is parallel efficiency. The code developed in this work was tested using the same mesh on various amounts of cpu cores to obtain a measure for the speed gained when using more cores. The most common performance metric for this is the speedup S, defined as:

$$S = \frac{T_1}{T_p} \tag{6.1}$$

Here T_1 is the execution time required using a single core and correspondingly T_p is the time required using p cores.

Also used in figure 6.0.2 below is the parallel efficiency E_p defined as:

$$E_p = \frac{T_1}{pT_p} \tag{6.2}$$

The test run was done using the mesh and setup from the square cylinder case (see section 5.3), performing 1000 time steps and recording the time required. The mesh consists of 962'560 cells and uses all the boundary types from the original case, including period ones (cyclicAMI). The old and the new pressure velocity scheme were investigated, as well as two types of linear solvers for the pressure equation: a multigrid solver (GAMG) and Krylov-based solver (PCG). Figure 6.0.1 shows the parallel speedup of all the cases. Table 6.0.1 provides the numerical values that were used in the figure.

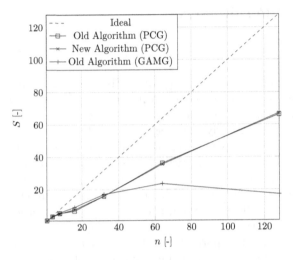

Figure 6.0.1: Speedup

No. of Cores	Cells/Core	T_o^{PCG} [s]	T_o^{GAMG} [s]	T_n^{PCG} [s]
1	962560	30326	34807	36831
4	240640	8749	9768	10680
8	120320	5506	5991	6987
16	60160	4385	3784	4774
32	30080	1892	2034	2294
64	15040	835	1472	1034
128	7520	458	2047	550

Table 6.0.1: Execution Times of Parallel Runs

It is evident that the GAMG solver is not suitable for massively parallel calculations. The speedup even starts to decrease below a certain number

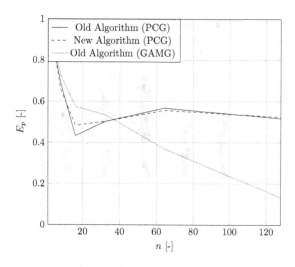

Figure 6.0.2: Parallel Efficiency

of cells per core. It could be observed that the number of iterations needed to solve the pressure equation increased with the amount of cores used, therefore slowing down the whole process. On the other hand, the PCG solver scales relatively well. Even super-linear scaling can be observed between 16 and 64 cores, ie. using twice as many cores reduces the time required by more than a factor of two (see figure 6.0.3 below). One thing to note is that the linear scaling (denoted as "ideal" in figure 6.0.1 above) can be attained by fully explicit algorithms where no equation systems need to be solved, ie. no matrix multiplications need to be performed. In our case, the pressure equation needs to be solved implicitly, therefore fully linear scaling (compared to the single-cpu effort) is out of reach.

The efficiency of the PCG solver decreases at the very last step, which leads to the conclusion that the optimal cell number per core is approximately 10'000 (when pure simulation speed is the goal). This value most likely depends on the setup of the computing cluster, however.

When comparing the old and new pressure-velocity coupling algorithms it can be seen that the new version is slightly slower, but scales just as well. This is due to an increase in molecule size in the pressure matrix, therefore

61

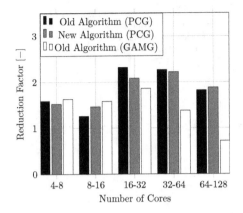

Figure 6.0.3: Time Reduction Factor When Doubling the Number of Cores

making the pressure equation more difficult to solve. The experiment using the GAMG solver with the new method leads to the same behaviour as above and is therefore not shown here.

7. Recommendations for LES Simulations

This chapter provides a few guidelines when dealing with large eddy simulations in general as well as specifically in OpenFOAM.

7.1 Discretisation Schemes

The discretisation practices successfully used during this work are given in the following table 7.1.1. The scheme names correspond to the names OpenFOAM uses. Other schemes might be possible for certain terms (such as the least squares practice for gradients) but have not been tested.

Term	Type
Time derivative	Euler or backward
Gradients	Gauss linear
Convection term	Gauss linear or Gauss limitedLinear 1.0
Diffusion term	Gauss linear
Laplacian terms	Gauss linear corrected
Surface normal gradients	corrected

Table 7.1.1: Discretisation Schemes

When using the solver developed in this work, the discretisation of the time derivative must be set to the Euler scheme. The third-order Runge Kutta scheme is hardcoded inside the solver and only works if this setting is chosen. The additional temperature scalar is also solved using this scheme and requires Euler as well. All other scalar transport equations, eg. used by the turbulence models, should be solved using the backward scheme (which is second order in time).

For the convection term of the momentum equations, only the Gauss linear scheme, which corresponds to central differencing, should be used for LES simulations. The central differencing scheme is non-dissipative and therefore allows an accurate representation of the rapidly changing flow field. Oscillations may occur as the scheme is unbounded, but these are not as critical. Any schemes that add artificial diffusion to keep the solution bounded are not recommended. The added diffusion may easily be of the same order as the turbulence model contribution and will strongly affect the solution. For additional scalars the limitedLinear 1.0 scheme has proven to be effective. Other schemes might be more suitable but this has not been tested.

7.2 Linear Solvers

The only linear system that needs to be solved when using the explicit solver is the pressure equation. As seen in the section on parallel performance on page 59 the PCG solver strongly outperforms the GAMG solver at high numbers of processor cores. In this work the PCG solver, in conjunction the FDIC preconditioner, has been used exclusively and performed very well. The pressure equation was generally solved to an absolute residual tolerance of 1e-6, which usually resulted in continuity errors lying in the range of 1e-12. The solver tolerance might be relaxed to speed up the simulation but the effect on the solutions has not been studied.

7.3 Initial Values

As far as initial values goes, it is generally advisable to start from a RANS solution. This solution does not have to be fully converged but should approximately represent the main flow field structures. If a RANS solution is not available and cannot be obtained it is advisable to use the perturbU utility in OpenFOAM to generate artificial, unstable turbulence structures. While this utility was developed primarily with fully developed channel flows in mind, it is also partially applicable to general cases. It is usually not feasible to start from constant solution fields as the fully turbulent state may take a very long time to develop.

7.4 Grid Resolution

The topic of grid resolution is widely discussed and apart from some general guidelines, no clear consensus is apparent in the literature as to what a proper LES-grid is. The preliminary work by Gaitonde [13] tries to address this issue and offers a methodology whereby the grid size required can be estimated a priori by evaluating the turbulent quantities of a RANS solution. The work mentioned above was published in 2008 and is part of a supposedly ongoing doctorate thesis, as the final results are not yet available.

One region where there seems to exist a general agreement on the mesh resolution requirements is near the wall. The dimensionless wall spacings (explained in section 5.2.4 on page 28) in the three coordinate directions should lie in the following regions (again see [13]):

$$x^+ \approx 35 - 55 \tag{7.1}$$

$$y^+ \leq 1 \tag{7.2}$$

$$z^+ \approx 15 \tag{7.3}$$

The paper by di Mare et al. [9] provides an overview of quality criteria that can be applied to large eddy simulations. According to the authors these quality indices should only be used as estimation tools and not as definite criteria. Two of these quality indices from this source are presented here as they are the most easily applicable in general cases (the original paper focuses strongly on the flow in internal combustion engines). These indices give an estimation for the amount of modelling present in the simulation rather than an actual quality value.

The first quality index IQ_ν is based on turbulent viscosity ν_t and is defined as follows:

$$IQ_\nu = \frac{\nu_t}{\nu + \nu_t} \tag{7.4}$$

Here a value of 0.5 would indicate that the turbulent viscosity is of the same magnitude as the laminar viscosity. To give an idea about the magnitude of these values IQ_ν has been calculated and averaged (based on the cell volumes) for the coarse and the fine channel case using the sigma subgrid model at the last timestep. For the coarse grid a value of 0.31 has been obtained while the fine grid gave a value of 0.18.

The second index shown here calculates a ratio between a characteristic mesh length scale and an effective Kolmogorov scale. This ratio determines how close the simulation is to a fully resolved DNS simulation.

$$IQ_\Delta = \frac{\Delta}{\eta_{eff}} \qquad (7.5)$$

Here Δ is the mesh length scale and simply equal to the filter width used in the model. The value for η_{eff} can be calculated using estimates for turbulent and numerical dissipation. The procedure is explained in [9].

A third and very promising approach to obtain a measure for the adequacy of the grid was suggested by Pope [34]. This approach was also used in an automatic grid refinement procedure in [28]. Pope introduces the following quantity:

$$M = \frac{k}{K + k} \qquad (7.6)$$

where

$$k = 0.5\langle \tau_r \rangle_T \qquad (7.7)$$

$$K = 0.5 \left(\langle \bar{u}^2 \rangle_T - \langle \bar{u} \rangle_T^2 \right) \qquad (7.8)$$

Here τ_r is the subgrid stress tensor, $\bar{u}$ is the resolved velocity and the angled brackets denote temporal averaging over a period of T. The quantity M gives an estimate of the ratio between modeled and resolved turbulence. A value of one would indicate full modelling (i.e. RANS) and a value of zero corresponds to a DNS simulation. A commonly accepted value for M in a LES simulation is 0.2 (see [18]), which results in about 80% resolution of the turbulent spectrum.

7.5 Mesh Quality

The following list is a selection from the Ercoftac Best Practice Guidelines [11] and is taken verbatim from [13]. It provides a good overview of general mesh quality aspects to be aware of. While anyone dealing with computational fluid dynamics should be aware of these points, they are given here as they are especially true in the context of large eddy simulations. Bad quality meshes, such as aspect ratios over 50 in the flow field, which are acceptable in stationary RANS simulations, can actually lead to solver failure.

1. Selection of a mesh global topology to help satisfy the specific code's requirements with regard to skewness, aspect ratio and expansion ratios.

2. Extent of the computational domain to be chosen to capture relevant flow and geometrical features. If required, the sensitivity of the calculation to the choice of computational domain to be examined.

3. Assessment of the geometrical features that can be omitted. In areas requiring fine local detail, considering local grid refinement. When employed, additional grid points should lie on the original geometry and not simply be a linear interpolation of more grid points on the original coarse grid.

4. Highly obstructed zones or fine detail of obstructions, if any, to be accounted for by the use of distributed losses or porosity.

5. Highly skewed cells to be avoided. For hexahedral cells or prisms the grid lines should be optimised in such a way that the included angles are approximately 90 degrees. Tetrahedra should tend to have their four angles equal.

6. Highly warped cells, that is cells with large deviations from co-planar faces to be avoided. Warp angles (measured between the surface normals of triangular parts of the faces) greater than 75 degrees can lead to serious convergence problems and deterioration in the results.

7. Non-orthogonal cells near boundaries to be avoided. This requirement is stronger than the requirement for the non-orthogonal cells away from the boundaries.

8. Tetrahedral elements to be avoided in boundary layers. Prismatic or hexahedral cells to be preferred because of their regular shape and ability to adjust in accordance with the near-wall turbulence model requirements.

9. High aspect ratios to be avoided in important regions of the flow domain but may be large in non-critical regions. This restriction may be relaxed near walls.

10. Specific code requirements for cell mesh stretching or expansion ratios to be observed. The change in mesh spacing should be continuous and mesh size discontinuities are to be avoided, particularly in regions of large changes.

11. Automatic grid adaptation techniques offered by some codes to be noted, as they might not always improve the grid quality (skewness, aspect ratio).

12. Critical regions with high flow gradients or with large changes (such as shocks, high shear, significant changes in geometry or where suggested by error estimators) to have a finer and a more regular mesh in comparison with non critical regions. Local mesh refinement to be employed in these regions in accordance with the selected turbulence wall modelling.

13. When using periodic boundary conditions, high geometric precision of the periodic grid interface to be ensured.

14. Arbitrary mesh coupling, non-matching cell faces, grid refinement interfaces or extended changes of element types to be avoided in the critical regions of high flow gradients.

15. Assumptions made when setting up the grid with regard to critical regions of high flow gradients and large changes to be checked with the result of the computation and grid points to be rearranged if found to be necessary.

16. Grid dependency study to be employed to analyse the suitability of the mesh and to give an estimate of the numerical error in the simulation.

8. Conclusions

In this work, the basic third-order time-accurate explicit solver for Open-FOAM previously developed in [19] was improved in such a way that it is now possible to simulate complex cases using the technique of large eddy simulation (LES). As a first step, the theoretical background of the LES procedure was explained. Afterwards, the turbulence models used in the context of this work are briefly described. More detailed information is given on the sigma subgrid model developed by [30], which was implemented from scratch as a part of the present study. A pressure-velocity coupling algorithm based on the works by Rhie and Chow [37] was implemented. It could be shown that the new algorithm effectively reduced pressure-oscillations at low timesteps and still achieved the predicted third-order accuracy in time.

The support for eddy-viscosity based LES turbulence models was added, with the limitation that models using additional scalar equations can only be treated second-order accurate in time. This has, however, not shown to be problematic for the quality of the results.

For the simulation of periodic cases, a switchable source term for the momentum equations was added that allows the prescription of a desired bulk velocity which is held upright by continuous correction of the flow field.

A temperature-transport equation was added for simulations involving heat transfer in incompressible flows. This additional scalar transport equation is solved in the same framework as the pressure and momentum equations, therefore allowing for a third-order accurate solution in time. Again, an artificial source term is available to deal with periodic flows involving heat flow in or out of the domain.

The solver was then validated on a series of test cases using measured data or data obtained by direct numerical simulations (DNS). The test cases were performed using a variety of different turbulence models described earlier.

The first case involved the simulation of the flow between two parallel plates. The results of averaged velocity and stress profiles were compared to DNS data by [25] and good agreement could be obtained, especially in the case of the fine grid.

Secondly, the flow around a square cylinder was investigated and aerodynamic data such as drag coefficients and Strouhal numbers were compared against measurements by [7] and simulation results by other authors. The sigma model produced good results in terms of averaged velocity profiles and drag coefficient, but all models predicted a Strouhal number that was approximately 4% higher than the measured value.

In the third case a periodic ribbed channel with heated walls was investigated and compared against measurements by [1]. Good agreement could be obtained for the Nusselt number profiles shapes, while the absolute values differed between the turbulence models. Very good results were achieved concerning the location of the reattachment point of the flow behind the rib.

The last case involved the simulation of a film cooling geometry where the results were compared against measurements conducted by [40]. The results were acceptable but the case needs further work. Good results are expected in the near future.

Additionally, a test case to assess the parallel performance of the solver was run. It was shown that the new pressure-velocity coupling algorithm was slightly slower than the old one but scales just as well. It could also be observed that the PCG linear solver using the FDIC preconditioner performed significantly better than the geometric-algebraic multigrid (GAMG) solver at high number of cpu cores.

Finally, based on the experiences gained in this work, recommendations for future simulations regarding numerical schemes, solvers, grid quality and grid resolution are given.

9. Unresolved Issues

During the course of this work a few issues have surfaced concerning solver stability. The implementation of the new pressure-velocity coupling scheme has delivered excellent results in the channel, square cylinder and ribbed channel cases but has lead to solver divergence in the film cooling simulation. The velocity values inside the cooling hole were increasing beyond the stability limit given by the CFL condition, therefore causing the simulation to abort. This phenomenon could be observed on different grids and even at very low timesteps which lead to the conclusion that it is not just a matter of stability but a bug in the code. The fact that it only occurred in the film cooling case suggests that one of the terms on the non-orthogonality correction of the pressure-velocity coupling is not working as intended. The film cooling case was the first and only case where the geometry did not allow for a fully orthogonal mesh, ie. the erroneous term was most likely zero in all the previous simulations. This will be investigated further.

Another source for instability was the use of time-varying boundary conditions, again a feature that was only used in the film cooling case. Here the problem was much less obvious and a simulation could run for several ten thousand timesteps and only crash later. The issue occurred when using the mapped boundary condition which copies values from inside the domain back onto the boundary, therefore creating a quasi-infinite channel at the inlet providing fully developed and physically meaningful turbulent boundary conditions. The divergence was observed using the new as well as the old pressure-velocity coupling method. As a first step the temporal scheme was reduced to a first-order Euler scheme and the problems remained. Another boundary condition creating randomised values around a prescribed mean was tested as well and also lead to instabilities. The discretisation scheme was then changed to first-order Euler implicit and the oscillations leading to the eventual failure disappeared completely. It is assumed that at the moment the boundary conditions are not updated to the proper values either before or after the solution of the pressure equation. At the moment it is unclear which solution variable causes the instability, but since the boundary condition for the pressure is a simple zero-gradient condition it is evident that either the velocity itself or the boundary flux field is wrong and needs to be treated specially.

Both these issues require attention, as both the treatment of non-orthogonal grids as well as time-varying boundary conditions are important for industrial flow problems.

10. Outlook

As a first step it is imperative that the problems described in the previous section are addressed.

Further, the ongoing discussion on mesh resolution will be followed, especially the results of the work by [13] as, if successful, they could provide a powerful tool to estimate the required LES grid size based on a RANS simulation. In that context, the meshing procedure described in the same document should be implemented in OpenFOAM. Similar meshes can already be produced by the SnappyHexMesh utility but the cells would not be able to have the staggered arrangement shown in [13], therefore only allowing coarsening factors of two from layer to layer. The BlockMesh utility could be rewritten but it would be limited to simple geometries. Ideally, a utility would be created that takes an existing hexahedral mesh and refines the near wall region in the appropriate fashion.

One topic that has been entirely omitted in this work is the use of hybrid RANS-LES models, due to the fact that all simulations here used meshes that fully resolved the wall region. For complex geometries and especially at high Reynolds numbers such methods could reduce the computational effort of time dependent simulations considerably. For accurate simulation results, the near wall resolution in wall normal direction would still have to be of the same order as in LES simulations but the resolution in streamwise and transverse direction could be relaxed dramatically. One of the most promising models of this type is the improved detached eddy simulation (IDDES) developed by Shur et al. [39]. It provides a framework for the blending of RANS and LES models and is implemented in OpenFOAM using the Spalart Allmaras (see [42]) formulations. The framework can be used with any existing RANS and LES models and could, eg. be coupled with the sigma LES model implemented in this work. The existing model of OpenFOAM has ben applied to the coarse channel mesh and the ribbed channel case. The results (see below) look promising but need further investigation. The Nusselt number profile is good, even if the absolute value is rather far away from the measurements. This will also have to do with the fact that the turbulent Prandtl number needs to be adjusted for hybrid models.

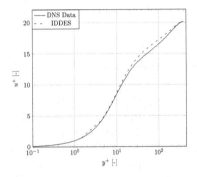

Figure 10.0.1: Mean Velocity Profiles, IDDES

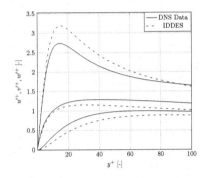

Figure 10.0.2: Mean Velocity Fluctuations, IDDES

As a final note, the implementation of stable spatial third-order procedure for unstructured grids would provide enormous benefits and should be the focus of future investigations. It would also give OpenFOAM an edge over commercial products and would probably increase its acceptance further.

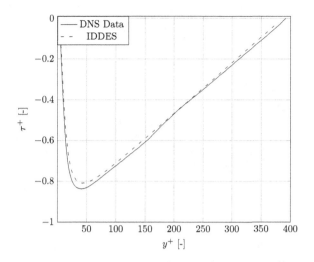

Figure 10.0.3: Mean Shear Stress, IDDES

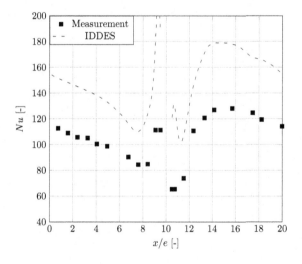

Figure 10.0.4: Nusselt Number, IDDES

Bibliography

[1] S. Acharya, S. Duttta, T. A. Myrum, and R. S. Baker. Periodically developed flow and heat transfer in a ribbed duct. *International Journal of Heat and Mass Transfer*, 36(8):2069–2082, 1992.

[2] Ansys Inc. *Fluent Users Guide*, 2006.

[3] A. Azzi and B.A. Jubran. Numerical modeling of film cooling from short length stream-wise injection holes. *Heat and Mass Transfer*, 39:345–353, 2003.

[4] Cosimo Bianchini. *http://www.cfd-online.com/Forums/openfoam/96885-compressible-wale-model.html#post377749*. University of Florence, 2012.

[5] J. Blazek. *Computational Fluid Dynamics: Principles and Applications*. Elsevier, England, 2001.

[6] F. Cadieux, G. Castiglioni, J. A. Domaradzki, T. Sayadi, S. Bose, M. Grilli, and S. Hickel. LES of separated flows at moderate reynolds numbers appropriate for turbine blades and unmanned aero vehicles. In *International Symposium on Turbulence and Shear Flow Phenomena*, Poitiers, France, 2013.

[7] S. Einav W. Rodi D. Lyn and J. Park. Laser doppler velocimetry study of ensemble-averaged characteristics of the turbulent near wake of a square cylinder. *Journal of Fluid Mechanics*, 304:285–319, 1995.

[8] Eugene de Villiers. *The Potential of Large Eddy Simulation for the Modeling of Wall Bounded Flows*. PhD thesis, Imperial College of Science, Technology and Medicine, 2006.

[9] Francesca di Mare, Robert Knappstein, and Michael Baumann. Application of LES-quality criteria to internal combustion engine flows. *Computers & Fluids*, 89:200–2013, 2014.

[10] E. R. Van Driest. On turbulent flow near a wall. *Journal of the Aeronautical Sciences*, 23(11):1007–1011, 1956.

[11] ERCOFTAC. Best practice guidelines. Technical report, Special Interest Group on Quality and Trust in Industrial CFD, 2000.

[12] J. H. Ferziger and M. Peric. *Computational Methods for Fluid Dynamics*. Springer, Germany, 2002.

[13] Ulka Gaitonde. Quality criteria for large eddy simulation, first year transfer report. Technical report, School of MACE University of Manchester, 2008.

[14] R. Henniger, D. Obrist, and L. Kleiser. High-order accurate solution of the incompressible Navier-Stokes equations on massively parallel computers. *Journal of Computational Physics*, 229:3543–3572, 2010.

[15] H. Jasak. *Error Analysis and Estimation for the Finite Volume Method with Applications to Fluid Flows*. PhD thesis, Imperial College of Science, Technology and Medicine, 1996.

[16] Blake Johnsson, Khai Zhang, Wei Tian, and Hui Hu. An experimental study of film cooling effectiveness by using PIV and PSP techniques. In *51st AIAA Aerospace Sciences Meeting including the New Horizons Forum and Aerospace Exposition*, Grapevine, Texas, 2013.

[17] W. Kim and S. Menon. Application of the localized dynamic subgrid-scale model to turbulent wall-bounded flows. Technical report, American Institute of Aeronautics and Astronautics, 1997.

[18] M. Klein. An attempt to assess the quality of large eddy simulations in the context of implicit filtering. *Flow, Turbulence and Combustion*, 75(1):131–147, 2005.

[19] David Roos Launchbury. Development of an explicit solver for incompressible flows in OpenFOAM. Semester work, University of Applied Sciences and Arts, Lucerne, 2014.

[20] A. Leonard. Energy cascade in large-eddy simulations of turbulent fluid flows. *Advances in Geophysics*, 18:237–248, 1974.

[21] D. Lyn and W. Rodi. The flapping shear layer formed by flow separation from the forward corner of a square cylinder. *Journal of Fluid Mechanics*, 267:353–376, 1994.

[22] W. Malalasekera and H. K. Versteeg. *An Introduction to Computational Fluid Dynamics: the Finite Volume Method.* Pearson Education, England, 2007.

[23] Charles Meneveau, Thomas S. Lund, and William H. Cabot. A lagrangian dynamic subgrid-scale model of turbulence. *Journal of Fluid Mechanics*, 319:353–385, 1996.

[24] P. Moin, K. Squires, W. Cabot, , and S. Lee. A dynamic subgrid-scale model for compressible turbulence and scalar transport. *Physics of Fluids A*, 3(11):2746–2757, 1991.

[25] Robert D. Moser, John Kim, and Nagi N. Mansour. Direct numerical simulation of turbulent channel flow up to $Re_\tau = 590$. *Physics of Fluids*, 11(4):943–945, 1999.

[26] Robert D. Moser, John Kim, and Nagi N. Mansour. $http://$ $turbulence.ices.utexas.edu/MKM\_1999.html$, 2007.

[27] S. Murakami and A. Mochida. On turbulent vortex shedding flow past a 2d square cylinder predicted by cfd. *Journal of Wind Engineering and Industrial Aerodynamics*, 54-55:191–211, 1995.

[28] A. Naudin, L. Vervisch, and P. Domingo. A turbulent-energy based mesh refinement procedure for large eddy simulation. Technical report, INSA - Rouen, 2007.

[29] F. Nicoud and F. Ducros. Subgrid-scale stress modelling based on the square of the velocity gradient tensor. *Flow, Turbulence, and Combustion*, 62(3):183–200, 1999.

[30] Franck Nicoud, Hubert Baya Toda, Olivier Cabrit, Sanjeeb Bose, and Jungil Lee. Using singular values to build a subgrid-scale model for large eddy simulations. *Physics of Fluids*, 23, 2011.

[31] K. Nozawa and T. Tamura. Les of flow past a square cylinder using embedded meshes. Technical report, Izumi Research Institute and Tokio Institute of Technology, Japan.

[32] J.S. Ochoa and N. Fueyo. Large eddy simulation of the flow past a square cylinder. Technical report, Area de Mecanica de Fluidos, Centro Politecnico Superior, Spain.

[33] Alberto Passalacqua. $https: // github. com/AlbertoPa$. Ames, Iowa, USA, 2014.

[34] S. B. Pope. Ten questions concerning the large-eddy simulation of turbulent flows. *New Journal of Physics*, 6(35), 2004.

[35] M. Porquie, M. Breuer, and W. Rodi. Computed test case: square cylinder. Technical report, Institute for hydromechanics, University of Karlsruhe, Germany.

[36] Osborne Reynolds. On the dynamical theory of incompressible viscous fluids and the determination of the criterion. *Philosophical Transactions of the Royal Society of London*, 186:123–164, 1895.

[37] C. M. Rhie and W. L. Chow. Numerical study of the turbulent flow past an airfoil with trailing edge separation. *AIAA Journal*, 21:1525–1532, 1983.

[38] P. J. Roache. Spectral methods for the navier–stokes equations with one infinite and two periodic directions. *Journal of Fluids Engineering*, 124(1):4–10, 2001.

[39] Mikhail L. Shur, Philippe R. Spalart, Mikhail Kh. Strelets, and Andrey K. Travin. A hybrid rans-les approach with delayed-des and wall-modelled les capabilities. *International Journal of Heat and Fluid Flow*, 29:1638–1649, 2008.

[40] A. K. Sinha, D. G. Bogard, and M. E. Crawford. Film-cooling effectiveness downstream of a single row of holes with variable density ratio. *ASME Journal of Turbomachinery*, 113:442–449, 1991.

[41] J. Smagorinsky. General circulation experiments with the primitive equations. *Mon. Weather Rev.*, 91:99, 1963.

[42] P. R Spalart and S. R. Allmaras. A one-equation turbulence model for aerodynamic flows. *AIAA Paper*, (92-0439), 1992.

[43] R. Verstappen and A. Veldman. Fourth-order DNS of flow past a square cylinder: First results. Technical report, Department of Mathematics, University of Groeningen, The Netherlands.

[44] Feng Wan, Fernando Porte-Agel, and Rob Stoll. Evaluation of dynamic subgrid-scale models in large-eddy simulations of neutral turbulent flow over a two-dimensional sinusoidal hill. Technical report, Saint Anthony Falls Laboratory, Department of Civil Engineering, University of Minnesota Twin Cities, Minneapolis, MN, USA, 2006.

[45] G. Wang and S.P. Vanka. LES of flow over a square cylinder. Technical report, Department of Mechanical and Industrial Engineering, University of Illinois at Urbana-Champaign, USA.

A. Appendix: Recent Developments

This section gives a brief overview of the activities performed after the initial report was handed in. These points were not part of the original thesis and are simply mentioned here for completeness.

The instability issue on non-orthogonal grids was fixed. The issue was that on such grids the normal boundary fluxes at wall boundaries was non-zero, therefore leading to conservation issues and ultimately simulation failure.

The second problem concerned time-varying boundary conditions causing instabilities. As suspected the old time values were the cause of the problem. The boundary values of the old time fields inside the Runge-Kutta cycle were never updated. This was not a problem in the case of cyclic, zero gradient or uniform fixed values, however, as soon as the fixed boundary values changed over time, the algorithm would fail.

The film cooling case was investigated further as well. The issues concerning the blowing rate could not be fully resolved due to the incompressibility assumption being inadequate. However, as the figures A.1 and A.2 below show, the prediction of the spreading of the cooling film downstream could be improved dramatically. For further case details please see chapter 5.5.

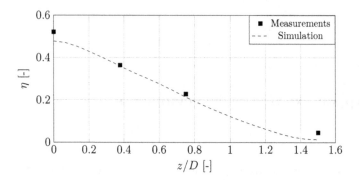

Figure A.1: Adiabatic Effectiveness at x/D = 6

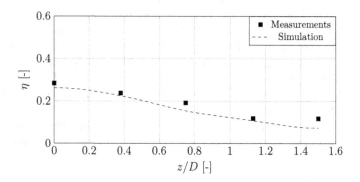

Figure A.2: Adiabatic Effectiveness at x/D = 15

Printed in the United States
By Bookmasters